THE
CONCISE
DICTIONARY
OF CRIME AND
JUSTICE

THE
CONCISE
DICTIONARY
OF CRIME AND
JUSTICE

MARK S. DAVIS

SAGE Publications
International Educational and Professional Publisher
Thousand Oaks ▪ London ▪ New Delhi

For information:

Sage Publications, Inc.
2455 Teller Road
Thousand Oaks, California 91320
E-mail: order@sagepub.com

Sage Publications Ltd.
6 Bonhill Street
London EC2A 4PU
United Kingdom

Sage Publications India Pvt. Ltd.
M-32 Market
Greater Kailash I
New Delhi 110 048 India

Printed in the United States of America

Library of Congress Cataloging-in-Publication Data

Davis, Mark S.
  The concise dictionary of crime and justice / Mark S. Davis.
    p. cm.
  ISBN 0-7619-2175-3 (hc)
  ISBN 0-7619-2176-1 (pb)
  1. Criminology-Dictionaries. 2. Criminal justice, Administration of-Dictionaries. I. Title: Crime and justice. II. Title.
  HV6017 .D38 2002
  364'.03-dc21

                      2002001200

This book is printed on acid-free paper.

02 03 04 05 10 9 8 7 6 5 4 3 2 1

Acquisitions Editor:	Jerry Westby
Editorial Assistant:	Vonessa Vondera
Copy Editor:	Dan Hays
Production Editor:	Denise Santoyo
Typesetter/Designer:	Larry K. Bramble
Cover Designer:	Michelle Lee

CONTENTS

PREFACE

Those working in criminal justice rely on a number of tools to make their jobs easier. Professors keep their lectures fresh by reading the latest research published in scholarly and professional journals. Researchers stay abreast of the latest statistical techniques. Practitioners search for "best practices" to help them better serve their clients. Regardless of whether the professional is a professor, a researcher, a practitioner, a journalist, or a crime writer, access to the vocabulary of criminal justice is a must.

There are several criminal justice dictionaries available. Some of these are directed more toward academic audiences, whereas others lean more toward the more popularized side of crime and justice. Some offer only the briefest definitions, whereas others are nearly encyclopedic in their coverage. All are excellent sources of terminology in the field.

So why yet another criminal justice dictionary? The dichotomy between, for example, scholar and crime writer and between professor and practitioner has narrowed in recent years. Many professors are former practitioners, and many practitioners have the credentials that would have enabled them to pursue academic careers. Because scholarship and criminal justice practice are more closely intertwined than ever, those engaged in these enterprises need easy access to the vocabulary of the field.

The Concise Dictionary of Crime and Justice is intended to be a concise guide to the vocabulary of contemporary criminal justice. Should readers need greater depth of information, they should consult reference works such as the several encyclopedias of crime and justice. There are also a number of Web sites that may offer more extensive treatment of selected topics.

ACKNOWLEDGMENTS

I thank Rolf Janke of Sage Publications, who agreed that there was room in the market for another criminal justice dictionary. I also acknowledge my profound appreciation to Jerry Westby at Sage, who, after inheriting the project, supported me with encouraging phone calls and e-mails and waited patiently for the final draft.

I recognize the substantial contributions of Rick Cooper in the genesis and development of this dictionary. Rick, a crime writer and armchair criminologist, initially was instrumental in identifying a number of the entries. Later, he took the time to review the entire manuscript, pointing out numerous oversights and factual errors. Every author should have a friend of his intelligence and devotion. He bears no responsibility for any remaining errors.

Finally, I thank my wife Jan. This project necessarily interfered with my participation in a number of family functions. Still, she withheld her disappointment and gently prodded me to complete the project as soon as possible, knowing there were many more waiting in the wings for completion.

abduction

the unlawful taking of a person by force, fraud, or persuasion. Abduction is similar to *kidnapping,* except that no demand for ransom is involved. An example of abduction is when a parent who does not have legal custody takes his or her child, keeping the child's whereabouts from the custodial parent.

abet

to encourage or assist an offender in the commission of a crime. See *aiding and abetting.*

abeyance

the state of a criminal sentence when it is suspended. When legal consequences are held in abeyance, the convicted offender generally must abide by certain conditions.

abolitionism

the movement to abolish the *death penalty* as a form of punishment. Abolitionism gained momentum in the late 20th century with the advent of *DNA testing* and its ability to exonerate the wrongfully convicted, including those awaiting execution. Abolitionists are particularly visible and vocal near the time of executions, often demonstrating outside a prison where an execution is about to take place.

abortion

the termination of a pregnancy. Criminal abortion is the act of intentionally producing a miscarriage or termination of pregnancy by any illegal means. Abortion became legal in the United States with the 1973 Supreme Court ruling in the case *Roe v. Wade.* Abortion remains a divisive issue, spawning acts of violence, including premeditated murder, against abortionists and abortion clinics by right-to-life extremists.

abrasion collar

a round hole with blackened margins made when a bullet pierces the skin. Abrasion collars not only indicate to investigators that the wound was made by a firearm but also may reveal the caliber, the distance between the shooter and victim, and other noteworthy forensic details.

Abscam

abbreviation for Arab Scam. Abscam was a scandal occurring from 1978 to 1980 in which U.S. government officials were caught on tape accepting bribes from *Federal Bureau of Investigation* agents posing as representatives of Arab sheiks. The officials, who included members of U.S. Congress, were convicted on a variety of charges. It was disconcerting to many that these public officials were so easily bribed. See *bribe, political crime.*

abscond

to leave a specific jurisdiction, especially one having legal control over an offender.

absolve

to release a person from any penalties, obligations, or consequences arising from a criminal act. Not necessarily a declaration of innocence.

Abt Associates

a Cambridge, Massachusetts-based research and consulting firm founded in 1965 by Dr. Clark Abt. Abt Associates, which uses research to address a variety of social problems, includes a law and policy area known as JUST. Abt Associates has undertaken a number of influential projects in criminal justice, including the development of a software package for neighborhood problem solving.

abuse

to injure physically, emotionally, or verbally. Abuse is also the result of such injury. See *child abuse, spousal abuse.*

abuse excuse
term used to describe the excuse that an offender's past abuse was responsible for the criminal conduct in question. The abuse excuse is used by some offenders and their attorneys as a *mitigating circumstance* of the alleged offense. Critics, including criminologist James Q. Wilson, argue that the abuse excuse is used too often to shirk responsibility for wrongdoing.

abuse of a corpse
the criminal misuse of a dead body. This offense can take many forms, from the improper handling or storage of the deceased by family members or funeral home personnel to acts of *necrophilia* by serial killers and other offenders.

abusive home
a home in which abuse occurs. The abuse can be physical, sexual, emotional, or verbal in nature; the victim of the abuse can be a person of any age. Abusive homes can pose a threat not only to the children who live there but also to successive generations of offspring affected by this influence.

Academy of Criminal Justice Sciences (ACJS)
an international membership organization that promotes professional and scholarly work in criminal justice. Established in 1963, the ACJS emphasizes criminal justice policy, education and research. *Justice Quarterly* is the scholarly journal of the ACJS, which holds its annual meetings in March of each year. The ACJS national office is located in Alexandria, Virginia.

Academy of Experimental Criminology
a scholarly body founded in 1999 to promote experimental design with random assignment in criminological and criminal justice research. Membership is by invitation and is based on the individual's contributions to experimental research in the field.

accessory
a person who, although not directly involved in the commission of a crime, assists the offender. See *accessory after the fact, accessory before the fact.*

accessory after the fact

a person serving as accessory after the actual commission of the crime. See *accessory, accessory before the fact.*

accessory before the fact

a person serving as accessory prior to the actual commission of a crime. An example of an accessory before the fact is someone who lent assistance to an offender prior to the actual event without actually participating in the crime. See *accessory, accessory after the fact.*

accidental killing

the unintentional taking of another's life. See *involuntary manslaughter.*

accomplice

a person who knowingly and willingly assists another person or persons in the commission or concealment of a criminal act. See *aiding and abetting.*

accusation

an allegation that a person has committed a crime. An accusation can be made without the initiation of formal charges.

accused

a person against whom criminal charges have been leveled. Compare with *suspect.*

accuser

one who brings an accusation against another.

acquaintance rape

the unlawful sexual violation of an individual by someone known to that person. The majority of sexual assaults against girls and unmarried women are by someone they know. See *date rape.*

acquit

to find a defendant not guilty by *jury, judge,* or panel of judges.

acquittal

the status of a criminal defendant of being found not guilty by a *jury, judge,* or panel of judges.

actus reus

literally, a wrong deed. Actus reus is an act that, combined with intent, constitutes a crime. See *intent, mens rea.*

ad curiam

at or to a court.

addiction

dependence on drugs, alcohol, or a certain habit. Addiction is composed of physiological dependence, psychological dependence, or both.

adipocere

a fatty, soap-like substance that can form when human or animal bodies decompose in moist, oxygen-deprived environments. Also referred to as grave wax and corpse fat.

adjudication

the process of disposing of a juvenile or criminal matter.

adjudicatory hearing

a formal court hearing at which a youth's case in juvenile court is disposed. At the adjudicatory hearing, the youth generally is placed on probation or, in cases of violent or otherwise serious delinquency, is sentenced to confinement in a correctional institution for youth.

admissibility

the status of a statement or evidence that permits it to be used in criminal proceedings against an accused offender.

admonish

to judicially warn. Judges and magistrates sometimes admonish those before the court for behavior that is disrespectful or otherwise unacceptable.

adult

generally, an individual who has reached his or her 18th birthday. With respect to criminal responsibility and consequences, adults are subjected to harsher penalties than their juvenile counterparts. Compare with *juvenile.*

adulteration

the modification of evidence so as to render it useless for prosecution purposes. An example of adulteration is the inadvertent contamination of evidence at crime scenes or in its subsequent handling so as to make it unanalyzable. One of the most famous criminal cases in which adulteration was alleged to have occurred was that of *O. J. Simpson*.

adultery

the act of a married person willfully engaging in sexual relations with someone other than the person's spouse. Although adultery was once punishable by death in many cultures and remains a crime in some jurisdictions, it has gained a certain degree of social acceptance.

adversarial justice

the system of justice in which there are two opposing parties. Western criminal justice processes are based on an adversarial system in which the prosecution and defense are pitted against one another. In civil cases, the adversaries are the plaintiff and the defendant.

adversary

the opposing party in a legal dispute.

affective violence

violence that is the result of highly charged emotions. An example of affective violence is when an individual, after a period of excessive drinking, reacts violently to an insult from another.

affidavit

a voluntary written statement of facts.

affirm

the act of validating an earlier decision or ruling. When an appellate court affirms a decision by a lower court, that decision stands.

affirmative defense

an acceptable rebuttal to a legal proscription against a certain type of behavior. For example, in many jurisdictions, an affirmative defense to the charge of carrying a concealed weapon is that the accused is a business person who regularly transports large sums of money and thus needs to carry a weapon for self-protection. An affirmative defense does not keep a person from being charged with a crime; if used successfully, however, it can result in dismissal of the charges.

affray
a fight between two or more persons in a public place causing a disturbance to others.

AFIS
see *Automated Fingerprint Identification System (AFIS)*.

aftercare
a period of supervised control of a releasee from a juvenile correctional facility. Aftercare is an opportunity to ensure compliance with special conditions, such as treatment or *restitution*. Compare with *parole*.

agent provocateur
a spy, often one who infiltrates an organization for the purpose of collecting information.

aggravated assault
a physical assault in which there is serious bodily injury or in which a weapon capable of inflicting such injury is used.

aggravating circumstance
a circumstance surrounding a crime that serves to increase its seriousness and the severity of the penalty. An example of an aggravating circumstance is the use of a firearm in the commission of a robbery.

aid panel
primarily in New South Wales, Australia: a group consisting of a police officer, a solicitor, community members, and young persons who work with the court in identifying opportunities for youthful offenders.

aiding and abetting
willingly and deliberately assisting another in the commission of a crime. See *accomplice*.

air piracy

the illegal commandeering of an aircraft by force or threat of force. Informally known as skyjacking. Certainly the most infamous example of air piracy is the September 11, 2001, takeover of American and United Airlines jets that were later crashed into the World Trade Center towers and the Pentagon. See *hijacking, piracy*.

AK-47

originally, a Soviet-made, fully automatic assault rifle and one of the most widely used weapons in the world. AK-47s are frequently mentioned in discussions of banning assault weapons. An AK-47 was used in the *mass murder* of schoolchildren in Stockton, California, in 1989. See *automatic weapon*.

al Qaeda

a Middle Eastern terrorist organization headed by *Osama bin Laden*. al Qaeda is thought to be behind a number of terrorist attacks against U.S. facilities, including the *Attack on America* on September 11, 2001. See *international terrorism, terrorism*.

alarm

a device that gives off a sound or signal intended to alert the occupants of a building or owner of a car to some violation or intrusion. See *burglar alarm*.

Alcatraz

an island and site of the former federal penitentiary in San Francisco Bay. Alcatraz, made infamous by some of its inmates, including *Al Capone*, George "Machine Gun" Kelly, Alvin Karpis of Ma Barker's gang, and Robert "The Birdman" Stroud, was used as a prison for difficult-to-manage offenders from 1934 to 1963. It was known as a prison from which escape was nearly impossible due to the strong currents and sharks in the bay. From 1969 to 1971, Alcatraz was occupied by Native Americans who were trying to reclaim Indian land and bring attention to the plight of the American Indian. Since 1973, Alcatraz has been a popular tourist attraction. See *federal prison, penitentiary*.

alcohol

ethyl alcohol, the intoxicating substance found in beer, wine, liquor, and other spirits. Alcohol has been linked in many ways to criminal behavior, including *victim-precipitated crime*.

alcohol and other drugs (AOD)

an umbrella term used to represent the general field of those who specialize in the prevention and control of substance abuse, including alcoholism.

alien

one who is not a citizen or legal resident of a country. See *criminal alien, illegal alien.*

allegation

a written or verbal statement claiming that someone has committed a crime.

allege

to make a formal claim that someone has committed a crime.

allocution

the right in common law of an individual to offer written or oral statements before a court. The right of allocution has been used as the legal basis for introducing *victim impact statements* and *privately commissioned presentence investigation reports* in court.

alpha

a statistical concept in criminological research that represents the probability that a person will reoffend. Alpha has played a significant role in research on the deterrent effects of various legal punishments. See *deterrence.*

altercation

a sometimes loud, angry dispute between two or more people often escalating to a physical attack.

alternative dispute resolution (ADR)

a set of methods and a movement to find ways of settling disputes outside traditional judicial processes. ADR includes the prevention of disputes and their resolution. See *night prosecutor's program.*

ambush

a surprise attack perpetrated by a person or persons unseen by the victim prior to the attack. Often, contract killings or mob "hits" take the form of an ambush, to catch the intended target off-guard.

American Academy of Forensic Sciences (AAFS)

a membership organization of physicians, criminalists, toxicologists, attorneys, document examiners, and others interested in forensic science. Headquartered in Colorado Springs, Colorado, the AAFS publishes the *Journal of Forensic Sciences* and holds its annual meetings each February.

American Bar Association (ABA)

a professional organization of more than 400,000 lawyers. Headquartered in Chicago, the ABA offers a variety of services, including law school accreditation, continuing legal education, and other programs designed to assist lawyers, judges, and other members of the legal profession.

American Bar Foundation (ABF)

the funding and research arm of the American Bar Association. The ABF has sponsored and conducted many influential sociolegal studies. Staffed by full-time employees and a number of visiting fellows, the ABF maintains a close working relationship and shares resources with Northwestern University and the University of Chicago. The ABF is headquartered in Chicago.

American Civil Liberties Union (ACLU)

a national organization dedicated to preserving the rights of individuals. Criminal justice issues in which the ACLU has been involved include *police brutality, racial profiling,* and the state of indigent defense. The ACLU frequently takes legal action against organizations whose activities threaten to endanger basic rights.

American Correctional Association (ACA)

a national organization of more than 20,000 correctional practitioners founded in 1870. The ACA, which holds an annual Congress of Corrections and a winter conference, is headquartered in Lanham, Maryland. Its publications include *Corrections Compendium* and *Corrections Today* magazine.

American Jail Association (AJA)

a national organization that supports those who operate and work in jails. The AJA, headquartered in Hagerstown, Maryland, publishes the magazine *American Jails.*

American Law Institute (ALI)

an organization of lawyers and legal scholars whose purpose is to promote the clarification and simplification of the law and its adaptation to social needs. Established in 1923, the ALI has an elected membership of 3,000 lawyers, judges, and law professors. At one time, the ALI actually took the time to define the law for professionals and laypersons alike, and it recently developed the *Model Penal Code*.

American Society of Criminology (ASC)

an organization formed in 1941 that supports and promotes criminology as a distinct professional field. The ASC is concerned with the entire spectrum of the criminal justice process and scholarly inquiry leading to new theory and knowledge. Each November, the ASC holds its annual meetings for its members, during which participants present papers and hold workshops. The ASC, which publishes *Criminology: An Interdisciplinary Journal* and a newsletter, *The Criminologist*, maintains its offices on the campus of Ohio State University in Columbus.

American Society for Industrial Security (ASIS)

an international organization of security professionals. Headquartered in Alexandria, Virginia, ASIS works to increase the effectiveness of security through educational programs and materials. It publishes the magazine *Security Management*.

amicus curiae

translation from Latin: friend of the court. A person with an interest in a matter before a court who files a brief in support of one of the parties.

amicus curiae brief

a brief filed by an *amicus curiae* on behalf of one of the parties.

ammunition

bullets, consisting of the projectile and its casing and propellant, used in a firearm. Also referred to as ammo.

amnesia

partial or total loss of memory, often arising as a result of trauma to the brain. Those accused of crimes have been known to claim amnesia as a defense; in other cases, the accused feigns amnesia.

amphetamines

any of several central nervous system stimulants. Active ingredients can include amphetamine, dextroamphetamine, or *methamphetamine*. Although amphetamines have many proper medical uses, such as medically supervised weight loss, they are also the drug of choice among many drug abusers. Street names include speed and *uppers*.

angel of death

a type of *serial killer*, often a nurse, medical technician, or other health professional, who believes he or she is relieving his or her victims of suffering by taking their lives. One of the most infamous angels of death was Donald Harvey, a Cincinnati area nurse who murdered a number of patients in his care.

animus

intention or motivation.

annoying calls

telephone calls made with the intent to threaten, harass, intimidate, or otherwise disturb someone. Teenagers often make such phone calls as a prank. Also referred to as prank calls.

anomie

a state of normlessness or lawlessness thought to be due in part to homogeneity in the population. See *anomie theory*.

anomie theory

a sociological theory, first formulated by French sociologist *Emile Durkheim* and later put forth by sociologist *Robert K. Merton* in *Social Structure and Anomie*, which posits that deviance occurs when there is an unequal emphasis in society on the ends people are expected to achieve and the means available to achieve them. Anomie theory is structural in that it attributes pathology such as crime to social forces rather than individual pathology. Later theoretical restatements include Richard Cloward and Lloyd Ohlin's *Delinquency and Opportunity*. Compare with *strain theory*.

anoxia

the deprivation of oxygen to the body. Significant brain damage or death can result if the period of anoxia is prolonged. Cerebral anoxia is a cause of death for those who commit suicide by hanging themselves.

ante mortem

occurring before death. This term is frequently used in autopsy reports to describe wounds or other conditions that occurred before a person died. Compare with *postmortem.*

Anti-Defamation League (ADL)

a national organization committed to exposing and fighting anti-Semitism and other forms of hatred and bigotry, including white supremacism and *Holocaust* denial. The ADL, founded in 1913, maintains regional offices throughout the United States and international offices. See *hate crime.*

antidote

any substance that counteracts the effects of a poison or other toxic substance.

antisocial behavior

behavior that does not conform to ordinary standards of decency or acceptability.

antisocial personality disorder

a personality disorder in which the person engages in violent, criminal, and other illegal and inappropriate behavior without remorse. Those suffering from antisocial personality disorder resist efforts at treatment. See *psychopath, sociopath.*

antitrust laws

federal and state laws designed to prevent price fixing and monopoly control.

AOD

see *alcohol and other drugs (AOD).*

Apalachin, New York

the site of a 1957 meeting of *organized crime* figures from around the country. Law enforcement authorities raided the estate where the meeting was being held, detaining the attendees and recording their license plate numbers. Apalachin served as evidence that there indeed was a *mafia* and that its tentacles reached across the entire United States.

APB

All points bulletin; a law enforcement announcement broadcast to authorities throughout a wide jurisdiction to be on the lookout for an offender, missing person, or other law enforcement emergency.

appeal

a postdisposition process in a criminal or civil case in which one of the parties formally argues in writing to higher courts that substantial mistakes were made at the lower court level.

appeal bond

a special bond that permits a convicted offender to remain free pending the outcome of an appeal. In the absence of an appeal bond, the offender would begin serving the sentence imposed by the lower court.

appeals court

a court higher than the court of original jurisdiction to which appeals are made and in which they are decided. Also referred to as appellate court. See *supreme court*.

appearance

the presence of a criminal defendant in court.

appearance bond

see *bond*.

appellant

the party that appeals to a higher court for a review of a lower court's decision.

appellee

the party in an appeal that argues the correctness of the lower court's decision.

appointed counsel

a private attorney appointed by the court to represent an indigent client. Appointed counsel often is used in jurisdictions not having a public defender. Some believe that the modest fees attorneys receive in such cases serve as a disincentive to the preparation of a strong defense. Compare with *public defender*.

apprehend
To capture, arrest, and take into custody a *suspect* in a crime.

arch
a segment of a human fingerprint. Compare with *loop, whorl.*

argot
a special vocabulary or slang unique to a group of people. Groups having their own argot include prison populations, street gangs, and those involved in the selling and using of drugs.

ARIMA
acronym for autoregressive integrated moving averages, a statistical technique for forecasting trends in data. ARIMA, which has been used to predict future crime rates and correctional populations, incorporates the lagged effects of variables on later events.

armed robbery
a robbery in which a weapon is used or feigned. Armed robbery is one of the more serious felonies. It is also known as aggravated robbery, in which the use of a weapon, most often a firearm, is the *aggravating circumstance.*

arraignment
the first appearance in the court of jurisdiction. During an arraignment, the defendant typically enters an initial plea, the court ensures representation by counsel, and bond is set. Compare with *initial appearance.*

arrest
the taking into custody of a person by police or other legal authorities with the intention of pressing criminal charges.

arrest clearance
the official removal by police of an active case following the arrest of a suspect. Because arrest clearances are a measure of police effectiveness, there is an incentive to clear arrests by any means, including through admission by offenders who may have had nothing to do with the crime in question.

arrest order

a written order issued by a *parole officer* or *probation officer* to arrest a parolee or probationer for a new offense or *technical violation* of rules.

arrest practices

the various means by which law enforcement officers effect arrests. Arrest practices can be controversial if they are thought to be abusive, discriminatory, or otherwise unfair or inappropriate. See *police brutality, racial profiling*.

arrestee

one who has been arrested. Compare with *detainee, suspect*.

Arrestee Drug Abuse Monitoring (ADAM)

a national effort sponsored by the U.S. Department of Justice to routinely collect data on drug use by those who are arrested and jailed. ADAM data show which drugs are being used by arrestees and which fall in and out of popularity. Trained data collectors interview arrestees and collect specimens at the various ADAM sites. ADAM includes males and females and both adults and juveniles. See *Drug Use Forecasting (DUF)*.

arsenic

a poisonous white powder at one time commonly used to commit murder. Arsenic is found in rat poison and certain herbicides.

arson

the malicious and unlawful burning of a building or other property. Arson includes the destruction of one's own property for *fraudulent insurance claims*. See *fire setting, serial arson*.

arsonist

an individual who commits arson. Compare with *fire setter*. See *arson, serial arsonist*.

art theft

the theft of fine art, such as paintings or sculptures.

Aryan Brotherhood

a white supremacist organization formed in San Quentin prison in 1967 to protect its members from blacks and Hispanics inside the institution. Now nationwide, members of the Aryan Brotherhood must commit murder to gain admission to the organization; membership ends only with the individual's death. Drug trafficking is a major source of income for the group.

Aryan Nations

a group of neo-Nazi extremists dedicated to the preservation of the white races.

asphyxia

the deprivation of oxygen leading to unconsciousness, injury, or death. Also referred to as suffocation.

asphyxiate

to suffocate or cause unconsciousness as a result of interference of the exchange of oxygen and carbon dioxide in the body.

aspiration of vomitus

the breathing of regurgitated food into the respiratory tract that cuts off the intake of air.

assailant

one who commits, or is suspected of committing, *assault*.

assassin

one who plans, attempts, or carries out an *assassination*.

assassination

the premeditated murder of a prominent person by surprise attack, often for political reasons. Infamous 20th-century assassinations include those of President John F. Kennedy, the Rev. Dr. Martin Luther King, Jr., and former Beatle John Lennon. The question of assassination as an appropriate course of retaliatory state action arose in the wake of the terrorist attacks on the World Trade Center and the Pentagon in September 2001. See *Lee Harvey Oswald.*

assault

the unlawful threat or touching of another person with intent to do bodily harm. Assault is often confused with *battery*. Also referred to as simple assault. See *aggravated assault, felonious assault.*

assault weapon

an automatic or semiautomatic firearm, generally with a large-capacity magazine, designed for firing a high volume of ammunition within a short period of time. An example of an assault weapon is the *AK-47*. There have been many legislative efforts to define assault weapons and to control their manufacture, distribution, ownership, and possession.

assembly line justice

a term used to convey justice processes so rote and routine that they compromise true justice. The notion of assembly line justice is reinforced by practices such as *plea bargaining* that, in many courts, occur in the vast majority of cases.

asset forfeiture

the legal requirement that certain accused or convicted offenders surrender real or other property believed to be fruits of their illegal activities. Asset forfeiture gained popularity in the 1980s with law enforcement agencies involved in the investigation of *drug trafficking*. Compare with *asset seizure*.

asset seizure

the taking by the government of money, property, or other items gained illegally through criminal activity.

asset-focused approach

the identification and use of an offender's positive assets, such as family and community support, as opposed to focusing on risks and deficits. Compare with *risk-focused approach*.

assisted suicide

the taking of one's own life with the help of another, often a physician or other medical professional. See *Jack Kevorkian*.

asylum

historically, a shelter such as a church or temple that offered protection from arrest. Also, the subject of a book, *Asylums*, by historian David Rothman.

atavism

a characteristic in an offender thought to be related to an earlier, more primitive form of being. For example, because a prominent brow was characteristic of Cro-Magnon, a similar feature on an offender might prompt some to believe that criminality is linked to less evolved forms of *Homo sapiens*. Nineteenth-century Italian criminal anthropologist *Cesare Lombroso* advanced the notion that certain types of offenders were throwbacks to an earlier form of evolutionary being and therefore could be identified by certain physical characteristics. See *Earnest Kretschmer*.

at-risk youth

a juvenile who by personal, family, community, or cultural characteristics is deemed vulnerable to engaging in deviant or delinquent behavior but who has not become involved in the juvenile justice system.

attachment

one of the four elements of Travis Hirschi's control theory of delinquency. Hirschi hypothesized that youths attached to their families and conventional values stand a greater chance of being insulated from delinquency. See also *belief, commitment, control theory, involvement*.

Attack on America

term used by politicians, the media, and others to describe the September 11, 2001, terrorist attacks on the World Trade Center, the Pentagon, and related terrorist plots against the United States. See *international terrorism, terrorism*.

attempted crime

a crime that has not been completed. Attempted crimes generally are punished slightly less severely than the corresponding *completed crime*.

attention center

same as *detention center*.

Attica

a state correctional facility in Attica, New York, that became a buzzword for prison reform after a bloody inmate riot in September 1971. Inmates, protesting crowded living conditions and possible racial overtones in inconsistent sentences and parole decisions, took over cell blocks for 4 days. The uprising ended when police stormed the facility and retook control in 15 minutes, with 39 inmates dying and more than 80 wounded in the process.

attorney general

the chief legal officer of the federal or state government. Attorneys general are appointed at the federal level but are often elected at the state level. Their responsibilities include representing the government in legal proceedings.

attorney-client privilege

the long-standing tradition of confidentiality that exists between attorneys and those they represent. Attorney-client privilege theoretically prevents the disclosure of information a client divulges to an attorney.

Auburn system

an outdated system of prison discipline characterized by segregation of prisoners in cells at night, walking in lockstep, maintaining silence, congregate work in shops at day, and dining with prisoners seated back to back while communicating with hand signals. The Auburn system was first employed at the Western State Penitentiary in Pennsylvania.

audit trail

a series of financial documents that, linked together, can support fiscal responsibility and, correlatively, uncover embezzlement, fraud, or other types of financial misconduct or crime. Audit trails are important in the investigation of *organized crime* and *white-collar crime*, whose perpetrators often go to great lengths to hide their illegally gotten assets through tangled webs of complex financial transactions.

Augustus, John

a Boston cobbler generally regarded as the father of *probation*. Beginning in 1841, Augustus supervised alcoholics and youth under an agreement with the local court.

auto theft

the theft of an unoccupied automobile, truck, or other similar vehicle. Auto theft is one of the major offenses that comprise the *Uniform Crime Reports* Crime Index. See *carjacking, unauthorized use of a motor vehicle.*

autoerotic fatality

a fatality that results from the practice of *autoeroticism.* Practitioners of autoeroticism often use elaborate props and restraints to carry out their sexual fantasies, most commonly with some type of hanging or neck compression to reduce the flow of oxygen into the body. Sexual gratification is achieved as the individual approaches unconsciousness. A self-rescue safety mechanism generally is incorporated into the practice, but in fatalities unconsciousness sets in before the escape method can be used. These deaths are often mistakenly deemed suicides or homicides by investigators unfamiliar with this phenomenon. In other cases, the true circumstances surrounding these deaths are kept from relatives to spare them emotional pain or public embarrassment.

autoeroticism

self-arousal and sexual satisfaction by means of fantasy or genital stimulation. See *autoerotic fatality*.

Automated Fingerprint Identification System (AFIS)

AFIS permits the electronic collection, storage, retrieval, and comparison of human fingerprints. These systems read, match, and store fingerprints. Although this does not eliminate the manual examination of fingerprints, it speeds up the process by reducing possible matches. See *fingerprint, LiveScan*.

automatic weapon

a firearm that fires continuously as long as the trigger is depressed or until the ammunition in the weapon is expended. Automatic weapons were used by the two bank robbers who terrorized North Hollywood, California, in February 1997. See *semiautomatic weapon*.

autopsy

a postmortem examination of the internal and external parts of the body to determine or confirm the cause of death. Same as postmortem.

autosadism

the infliction of pain on one's self for sexual gratification. Compare with *sadism*.

aversion therapy

a form of psychodynamic intervention in which the patient is subjected to an unpleasant sensation such as electric shock in conjunction with an image to be extinguished. For example, pedophiles have been treated with aversion therapy by viewing pictures of children while receiving a mild electric shock, theoretically conditioning them to thereafter associate sex with children with pain. Also referred to as aversive conditioning.

avulsion

the tearing away of a body part or tissue as a result of trauma or a surgical procedure. The examination of avulsions by medical personnel or forensic investigators can often point to the kind of weapon used in a violent crime.

baby boom

a noticeable increase in the birthrate within a relatively short period of time. Usually, this term refers to the post-World War II population explosion of children fathered by returning veterans. Baby booms carry with them higher numbers of persons, which causes increases in crime when the "baby boomers" reach offending ages. The same phenomenon will cause problems in correctional populations as confined offenders age and suffer health and other age-related problems. This bulge in the population has also been descriptively referred to as a "pig in a python." See *echo boom*.

backlog

in a criminal court, the cases awaiting disposition. A number of solutions have been sought for the problem of case backlogs, including *diversion*. See *caseload reduction, court delay, speedy trial*.

bad check

a check written with insufficient funds in the account to cover it. The writing of a bad check can be unintentional or intentional, the latter sometimes being defined as *passing bad checks*.

bad seed hypothesis

a theory that certain human beings are born with the capacity to engage in criminal behavior. Once dismissed, this hypothesis has more credence with the advent of genetic evidence that suggests the inheritability of criminal traits. See *biocriminology*.

bad time

days added to the sentence of an imprisoned offender for poor conduct. Bad time became popular with the advent of *determinate sentencing* and the more punitive philosophy of *retribution* of the 1970s and 1980s. Also bad time credit. Compare with *good time*.

badgering

the harassment of a *witness* giving testimony in a court of law. This tactic is often employed in order to make strong impressions on the jury or to confuse a witness into making contradictory statements.

bagging of hands

the tying of paper bags around the hands of homicide victims in order to preserve evidence. If the victim struggled with the attacker, bits of skin and blood may be found under the victim's fingernails. Paper bags are used because plastic bags cut off the air flow, increasing the rate of decomposition and potentially changing the chemical composition of any trace evidence that might be present.

bail

money, property, or other security offered in exchange for the release from custody of an arrested person and to guarantee the person's appearance at trial. Bail is forfeited if the accused does not appear in court. See *bond, Manhattan Bail Project, pretrial release.*

bail agent

a person who finds and takes into custody those who have skipped out on bail bonds. Although the tactics of bail agents have been surrounded by controversy, higher courts have upheld the constitutionality of their work. Some states, most notably California, mandate standardized training for those who want to serve as bail agents.

bail bondsman

a person who makes a living ensuring the *appearance* of criminal defendants in return for a fee, often a percentage of the bail bond set by the court. Bail bondsmen are supposed to forfeit to the court the bond if the defendant fails to appear, but in practice this seldom happens. See *surety.*

bail reform

the movement to make the practice of bail more equitable for the accused. Studies have shown that bail practices of the past discriminated against certain segments of the population, including minorities and the poor. See *Manhattan Bail Project, pretrial release.*

bailiff

an officer of the court whose job entails controlling access to the judge or magistrate, keeping the judge's calendar, and maintaining order in the courtroom. Generally, bailiffs are appointed by the judges for whom they work.

bait and switch

a deceptive practice used by some retailers who try to persuade a customer to buy a different, often more expensive item rather than the advertised item that drew the customer to the store. This practice is illegal if the advertised item was never actually available. See *false advertising.*

ballistics

the scientific study of projectiles either still in the bore of the firearm or after the weapon is fired. Microscopic examination of the markings on a spent bullet can determine if a specific weapon fired it. Ballistics is also a term used to describe the labs or units that conduct studies of projectiles, their trajectories, and the firearms from which they are fired.

banditry

crime, often robbery and sometimes including murder, committed by outlaw bands.

banishment

a punishment in which the offender is forced to leave his or her original homeland, sometimes to a specified alternative location for a specified period of time. Banishment was popular in the 18th and 19th centuries in Europe. It has more limited value in modern times because many nations now refuse to accept convicted offenders from other counties. Compare with *transportation.*

bar

the court. Also, all the judges, prosecutors, attorneys, and other legal practitioners who participate in the local justice system.

Barbie, Klaus

(1913-1991) a former Nazi Gestapo commander during World War II who was charged with committing atrocities against Jews. Barbie was known as "the butcher of Lyons." See *holocaust, war crimes.*

barbiturates

a group of sedative drugs derived from barbituric acid. Physiologic effects include decreased blood pressure, respiration, and heart rate.

Barnes, Harry Elmer

(1889-1968) a historian of the early to mid-20th century who chronicled the history of confinement, torture, and execution in his book, *The Story of Punishment*.

barrister

in England, a counselor who has been admitted to the *bar* and thus can advocate for clients. Compare with *counselor, solicitor*.

barroom violence

violence that occurs in or around a bar, tavern, or other establishment where *alcohol* is sold for consumption on the premises. Barroom violence, studied as early as the 19th century by Belgian statistician *Adolphe Quetelet*, is fueled by the intoxication of the parties and facilitated by the carrying and use of weapons.

baton

a long stick fashioned of wood or synthetic material designed to permit law enforcement officers to bring suspects under control without the use of *lethal force*. See *PR-24, nightstick, billy club, truncheon*.

battered child syndrome

a psychological condition brought on by a pattern of behavior that develops in children who are subjected to parental abuse that occurs regularly over time, resulting in the conclusion that violence against the parents or caregivers is the only way to end the abuse.

battered wife syndrome

a psychological condition brought on by a pattern of behavior that develops in women who are subjected to a form of spousal abuse that occurs regularly over time, resulting in the conclusion that violence against the husband is the only way to end the abuse.

battery

intentional physical contact with another person without that person's consent. Laypersons often confuse battery with *assault*.

bawdy house
a house of prostitution. Also referred to as a brothel, a house of ill repute, and a whorehouse.

beat
the geographical area routinely patrolled by a law enforcement officer.

Beccaria, Cesare Bonesa, la Marchese di
(1738-1794) an Italian nobleman and intellectual of the late 18th and early 19th centuries who wrote a famous essay, *On Crimes and Punishment*. In his book, Beccaria spoke out against a number of unjust practices of the day, including *torture*, secret accusations, the abuse of power by judges, and inconsistencies in sentencing offenders. Beccaria's book was a success in Europe, appearing at a time when intellectuals were questioning a number of social institutions of the day. See *classical school of criminology*.

bed-wetting
the involuntary enuresis of children or adolescents. Bed-wetting is considered one of the three early indicators of *psychopathy*, the other two being *fire setting* and *cruelty to animals*.

behavior modification
the process of altering the maladaptive behavior of individuals through the use of classical conditioning and operant conditioning. See *differential association-reinforcement theory*.

Behavioral Science Unit (BSU)
the unit of the *Federal Bureau of Investigation (FBI)* that uses in-depth crime scene analyses, criminal personality profiling, and other techniques to solve homicides and other violent and serial crimes. Its mission includes the development and provision of training, research, and consultation programs, grounded in the social and behavioral sciences. What makes the BSU unique is its emphasis on studying not only offenders but also their motivation and behavior. The BSU is headquartered at the FBI National Academy in Quantico, Virginia. See *Child Abduction Serial Killer Unit (CASKU)*, *National Center for the Analysis of Violent Crime (NCAVC)*.

belief

one of four elements of Travis Hirschi's control theory of delinquency. See *attachment, commitment, control theory, involvement.*

belly chain

a steel chain designed to pass through the belt loops of an individual in custody, to which handcuffs are attached. This prohibits the individual so constrained from using the arms as weapons. See *handcuffs, leg irons, shackles.*

bench trial

a trial in which the verdict is rendered by the presiding judge. In bench trials, the defendant must waive his or her constitutional right to trial by jury. Bench trials can be presided over by a single judge or, in some cases, by a three-judge panel.

Bentham, Jeremy

(1748-1832) a jurist and philosopher of the late 18th and early 19th centuries. Bentham is best known for his writings that advocated for the abolition of the death penalty and the imposition of punishments commensurate to the seriousness of the crime. He cast man as hedonistic but who could freely choose among various courses of action, including those defined as criminal. See *classical school of criminology.*

Berkowitz, David

see *Son of Sam.*

Bertillon, Alphonse

(1853-1914) a French police records clerk who developed the *Bertillon classification system.*

Bertillon classification system

a now archaic system of using various anthropometric measurements to identify and categorize criminals, named for its developer *Alphonse Bertillon.* Consistent with prevailing criminological thought of the day, the Bertillon system consisted of measurements of the head and body to classify offenders. It was eventually discredited due to its inability to ensure uniform measurement and by the advent of fingerprinting, which was proven to have greater reliability. See *criminal anthropology, Cesare Lombroso, positive school of criminology.*

bestiality

sexual acts between a human being and an animal. An individual charged with bestiality would likely be charged with cruelty to animals. Also referred to as *zoophilia*.

betting

wagering or *gambling* on any of a number of activities, including sporting events, horse and dog races, card and dice games, or other games and activities governed by skill, chance, or both.

bifurcated process

a two-part trial. An example of a bifurcated process is a capital trial in which the first part centers on a finding of guilty or not guilty and the second part generally has the jury deciding between *life imprisonment* or the *death penalty*.

bigamy

the criminal offense of being legally married to more than one spouse at the same time. Despite its illegality, bigamy continues to be practiced by some Mormons and others. There are instances of men who have maintained two families at the same time.

bill of attainder

the condemnation of person without a formal trial. Bills of attainder are prohibited by the U.S. Constitution.

bill of indictment

see *indictment*.

bill of information

a charging document occasionally used by prosecutors when a felony case is not taken to a grand jury. Bills of information are sometimes used by prosecutors to charge those who are not going to contest the charges.

bill of particulars

a document citing specific allegations against a person.

Bill of Rights

the first 10 amendments of the U.S. Constitution. The Bill of Rights restricts the role of federal government by guaranteeing a number of individual freedoms, including protections for those suspected or accused of committing crimes and the right to bear arms.

billy club

a rod-like weapon, often fashioned from hardwood, used by law enforcement officers to bring arrestees under control. See *baton, nightstick, PR-24, truncheon.*

bin Laden, Osama

a terrorist leader considered responsible for a number of terrorist acts, including the World Trade Center and Pentagon attacks in 2001. Osama bin Laden, born into a wealthy Saudi Arabian family, renounced his Saudi citizenship, was disowned by his family, and thereafter committed himself to fighting the perceived Western enemies of Islam. See *al Qaeda, Attack on America.*

binge drinking

the practice, often by college students, of drinking large amounts of alcoholic beverages in a short period of time. Binge drinking has been defined as four or more alcoholic drinks within 1 hour. Although binge drinking is not in and of itself a crime, those who binge drink frequently engage in conduct that is criminal, including *assault, vandalism*, and *public indecency.*

biocriminology

the consideration of biological factors in the etiology of criminal behavior. See *twin studies.*

biological determinism

the belief that certain biological factors are responsible for the genesis of criminal behavior.

biological evidence

evidence that derives from the human body, such as blood or tissue. Biological evidence permits officials to employ *DNA testing* to link offenders to specific crimes. Even in cases in which bodies are severely burned, biological evidence can sometimes be obtained from bone marrow or the pulp of teeth.

bioterrorism

the use of biological material, such as anthrax or botulin, to perpetrate *terrorism*. Long a potential threat, bioterrorism became more salient in the wake of the *Attack on America.*

birth cohort

a group of individuals who have their year of birth in common. In a well-known longitudinal study, criminologists *Thorsten Sellin* and *Marvin E. Wolfgang* tracked the criminal careers of a cohort of males born in 1945 to determine how many went on to become involved in serious delinquency.

birth order

the order in which a person is born in relationship to his or her siblings. Birth order has been studied by criminologists and investigators as one of many possible factors involved in the genesis and production of criminality.

bite-mark identification

identification of a suspect through the impression of tooth marks left in the skin of the victim or in other materials, such as chewing gum or food. Often used in conjunction with saliva washings of the bite-mark area on the skin, which can yield blood groups on serological examination. Bite-mark identification has come under attack due to the lack of scientific consensus on this specialized field of study.

Black Hand

an extortionist group active in the United States in the early 20th century. The Black Hand had its origins in Sicily. Also referred to as Mano Nera.

black market

the illegal trafficking in merchandise that is either in scarce supply or whose manufacture or distribution is heavily regulated.

Black Panther Party

a group founded in 1966 by Bobby Seale and Huey P. Newton. The Black Panthers, known for advocating the end of racial discrimination and economic oppression of blacks, were also involved in food giveaways and free health clinics.

Black Talon

a special bullet, outlawed in many jurisdictions, that causes maximum damage to human tissue due to sharp points that extend on impact. Compare with *cop-killer bullets*.

blackmail

the illegal act of demanding payment or other benefit from someone under the threat of physical or other harm if payment is withheld. Compare with *extortion*.

blended sentencing

a mixture of juvenile and adult sentences for juvenile offenders. In blended sentencing, a juvenile may be sentenced to serve time in a juvenile institution but may later be transferred to an adult institution upon reaching age 18. Blended sentencing became popular in the 1990s during the movement for tougher measures for juveniles.

block watch

a group of neighbors in housing subdivisions or other cluster-type developments who join forces for the purpose of watching for and reporting suspicious people or activities in their neighborhood. Police departments often assist in forming and advising block watch groups. Signs are posted prominently on the streets in the neighborhood to warn criminals that their activities might be observed by local block watch members. See *crime prevention*.

Bloods

a notorious Los Angeles street *gang* that has spread throughout the United States. See *Crips, Latin Kings*.

bloodstain

the residue of blood. Bloodstains reveal numerous clues to crime investigators, including not only blood type but also the position of the victim and other details of the crime.

bludgeon

to strike with a heavy club or club-like object. Also, a club or object so employed.

bluebeard

a term given to a man who serially marries and then murders his wives, based on a fictional character in a story by Charles Perrault.

body armor

bullet-proof vests and other clothing designed to stop the penetration or lessen the impact of bullets. See *bullet-proof vest, Kevlar*.

Boesky, Ivan

an infamous Wall Street *insider trader* of the 1980s. Boesky used confidential stock information to illegally gain millions of dollars. In return for leniency, he agreed to help authorities convict *Michael Milken* and others involved in the same scheme.

bogus

phony; not genuine. Bogus is often used to describe counterfeit money or credentials such as driver's licenses.

boiler room fraud

a scam in which the perpetrators work out of rented offices known as boiler rooms. Using banks of telephones, they try to solicit donations for nonexistent charities or interest potential investors in fraudulent companies. See *telemarketing fraud*.

bomb squad

a unit of a law enforcement agency specially trained to safely handle, move, disarm, and detonate bombs and explosives.

bond

a written guarantee of performance. See *bail*.

bond agent

see *bail bondsman*.

bondage

masochistic activities generally involving bindings and restraints, hoods, gags, and blindfolds. Bondage can be practiced with a single partner, in groups, or alone. Sometimes, bondage is taken too far, resulting in serious injury or death. See *autoeroticism, sadomasochism*.

bonding

see *bond*.

bondsman

see *bail bondsman*.

bookie

a bookmaker; a person who takes illegal bets.

booking

the recording in official police records of the facts pertaining to an individual's arrest, including identifying information and the specific charges that were filed. It is common for a suspect to be fingerprinted as part of the booking process.

bookmaking

the illegal taking of bets.

booster

a thief, particularly a type of professional shoplifter. Also *snitch*.

booster girdle

an elastic band or similar device, worn around the waist or other part of a shoplifter, for the purpose of holding stolen goods tightly against the body to conceal them.

boot camp

a correctional facility for youthful or first-time offenders intended to operate similarly to a military boot camp. Boot camps use structure, discipline, physical training, and verbal intimidation by drill instructors to break down and rebuild offenders. Research has shown that boot camps, although effective in helping offenders in the short term, have not been as successful in obtaining long-term changes in behavior for many offenders.

bootlegger

a person who engages in the illegal manufacture, sale, or transportation of *alcohol*. The term is derived from the practice of *smuggling* alcohol by concealing it in an actual bootleg. Bootlegging was prevalent during the U.S. experience with *Prohibition*. See *Bureau of Alcohol, Tobacco and Firearms (BATF)*, *Volstead Act*.

Borden, Lizzie

a Fall River, Massachusetts, woman who is alleged to have viciously murdered her own parents with a hatchet in 1892. Borden was tried and found not guilty by a *jury*.

Border Patrol, United States

a division of the U.S. *Immigration and Naturalization Service* responsible for preventing the entry of illegal aliens and smugglers into the United States. The Border Patrol employs a variety of means to detect the passage of people and vehicles, including electronic surveillance and air, marine, and horse patrols. In addition to apprehending illegal aliens, the Border Patrol routinely seizes substantial amounts of illegal drugs being smuggled into the United States.

borderline personality disorder

a personality disorder characterized by instability of relationships and impulsive behavior. Those with borderline personality disorder have an unstable self-image and may engage in self-mutilation and suicidal behavior. Borderline personality disorder is a common diagnosis among confined female offenders.

borstal

a residential facility in Britain in which youthful offenders learn a trade such as masonry or carpentry, receive an education, and participate in counseling. Youthful offenders spend from 6 months to 2 years at borstal. Dating back to the early 1900s, borstal takes its name from Borstal, England, where the first such facility was opened.

Boston Strangler

the nickname given by the media to Albert DeSalvo, a *serial killer* who is alleged to have murdered at least 13 women in the Boston area from 1962 to 1964. Despite DeSalvo's confession to 11 of the murders, there are many who do not believe he was the true Boston Strangler. DeSalvo eventually was killed by another inmate in prison.

botanical material

fragments of plants that can be used in solving crimes. Botanical material may be found on a suspect's shoes, in the tire treads of vehicles, or in other obscure places. For example, the analysis of botanical material can lead investigators to the original crime scene in cases in which the offender transported the body after the crime.

bounty hunter

a privately employed agent who tracks down and apprehends wanted and fleeing felons in return for reward money. Bounty hunters have broad powers under the U.S. Constitution, and, as such, are not regulated. Consequently, they have been at the center of much controversy over some of the questionable tactics they use in apprehending felons. See *bail agent*.

Brady Bill

a piece of federal legislation named after former White House press secretary James Brady, who was shot and disabled in the 1981 *assassination* attempt on President Ronald Reagan. The Brady Bill was designed to reduce the opportunities for criminals to obtain handguns by requiring purchasers of handguns to undergo a criminal background check. See *Handgun Control, Inc., National Instant Check System*.

Branch Davidians

a group who lived with and followed the teachings of David Koresh at a complex near Waco, Texas. Koresh and a number of Branch Davidians died in a shootout with federal authorities in 1993.

brass knuckles

a hand-held device, fashioned from brass rings or other cast metals, used to inflict greater harm when hitting a person with the fist.

Brawner test

the legal test of mental illness that asserts that a person is not responsible for a crime if at the time of the crime the conduct was the result of mental disease or defect such that either the individual is incapable of appreciating the wrongfulness of his or her conduct, or he or she cannot conform his or her conduct to what is required by the law. See *insanity defense*.

break-in

the forced entry into a building for the purpose of committing a crime.

breaking and entering

the felonious entering of a dwelling or business for the purpose of committing theft. See *burglary, criminal trespass*.

Breathalyzer

a machine that measures the level of alcohol in a person's bloodstream. Breathalyzers are used by law enforcement officers to test the blood alcohol content of *driving while intoxicated* suspects. The accused breathes air through a plastic tube that is analyzed by an alcohol simulator. A digital readout displays the results of the measurement.

bribe

money given illegally to another in return for a favor or consideration.

bribery

the act of giving, offering, or accepting a bribe.

bridewell house

in England, a prison.

brief

a legal document that asserts a position, poses a legal question, or makes a request.

British Society of Criminology

the professional membership organization of criminologists in the United Kingdom. The British Society of Criminology, which has a membership of more than 800, holds an annual conference and publishes the highly regarded *British Journal of Criminology*.

broken windows

a phrase taken from an influential and oft-cited 1982 article in *The Atlantic Monthly* by criminologist George Kelling and political scientist James Q. Wilson. Kelling and Wilson asserted that outward manifestations of community decay, such as broken windows and unmown vacant lots, lead to more serious decay of urban neighborhoods, creating fear in residents and opening the door to violence and other serious criminal behavior.

broken windows probation

a form of probation supervision grounded in the *broken windows* theory that helped spawn *community policing*. Broken windows probation connects probation more closely with the problems and needs of the communities in which probationers live and are supervised, focusing on programming that works.

brothel

a house or other structure in which prostitution takes place. Also *bawdy house*, house of prostitution.

buggery

an unnatural sex act between humans or between a human and an animal. Compare with *bestiality, sodomy, zoophilia*.

bullet jacket

a sheath, usually made of copper, that encases a lead bullet for the purpose of reducing lead fouling of the inside of the gun barrel. Bullet jackets can be analyzed to determine characteristics of the bullet and the gun from which it was fired.

bullet track

the path made by a bullet entering and passing through the body. Examination of bullet tracks can yield useful information about the position and angle of the weapon when fired and the relative positions of the perpetrator and victim. See *ballistics, trajectory*.

bullet-proof vest

a vest that is impenetrable by bullets. Worn by many police officers but also utilized by criminals. With the application of *Kevlar*, bullet-proof vests became much more effective in providing protection for the wearer. This, in turn, was compromised by the use of Teflon-coated bullets, which can pass through such vests. Consequently, some bullet-proof vests incorporate impenetrable steel plates. See *body armor*.

bull pen

a holding area in a jail or lockup for those awaiting arraignment or transfer to another facility.

bullying

threatening or intimidating behavior, sometimes accompanied by physical violence, by some youths toward others, especially those smaller, weaker, or otherwise perceived to be vulnerable. Bullying has been the target of specially tailored prevention and intervention programs.

Bundy, Theodore "Ted"

(1946-1989) a prolific serial killer of the 1970s. Bundy abducted and murdered numerous girls and young women in the states of Washington, Utah, and Colorado. The former law student eventually was apprehended after he murdered two young women in a sorority house at Florida State University. He was convicted of these crimes, along with the rape and murder of a 12-year-old girl, and was executed in Florida's electric chair in 1989. Throngs of people celebrated his execution outside the prison at the time of his death.

burden of proof

the obligation on the part of the government to prove that an accused person is guilty of committing a crime. In the United States, such proof in a criminal case must be beyond a reasonable doubt, a standard higher than that required in civil proceedings.

Bureau of Alcohol, Tobacco and Firearms (BATF)

a organizational unit of the U.S. Department of the Treasury responsible for the enforcement of federal laws pertaining to the manufacture and distribution of alcoholic beverages, tobacco products, and firearms. Its responsibilities encompass the investigation of bombings such as that of the World Trade Center in 1993. The BATF drew much criticism for its role in storming the *Branch Davidians'* complex in Waco, Texas, on February 28, 1993, during which a number of men, women, and children died.

Bureau of Justice Assistance (BJA)

a subdivision of the *Office of Justice Programs (OJP)*, U.S. Department of Justice. BJA is responsible for administering formula and discretionary grant programs designed to prevent and control violent and other crime and to improve the administration of criminal justice. Among the grant programs administered by BJA is the Edward Byrne Memorial State and Local Law Enforcement Assistance Program, named for a New York police officer killed in the line of duty.

Bureau of Justice Statistics (BJS)

the branch of the *Office of Justice Programs (OJP)* within the U.S. Department of Justice that promotes the collection and analysis of crime data in the states and territories. BJS accomplishes this in part by awarding funds to state units known as *statistical analysis centers*. BJS has been instrumental in promoting the *National Incident Based Reporting System (NIBRS)*.

burglar

a person who commits burglary. See *cat burglar*.

burglar alarm

an electronic device designed to alert residents or car owners that a *break-in* has occurred. Burglar alarms use a variety of technologies, most recently cellular telephones, to notify authorities of an intrusion.

burglary

the unlawful entry into an unoccupied dwelling or vehicle for the purpose of committing theft. Laymen often confuse burglary with *robbery*. Compare with *breaking and entering, criminal trespass*.

burking

murder by asphyxia and smothering in such a way as to disguise the manner of death by leaving the victim unmarked. Some cases of sudden infant death syndrome are suspected of actually being instances of burking.

business crime

crime committed in the course of transacting business. Types of business crimes include *white-collar crime* and *corporate crime*.

bust

an arrest.

butt

the grip end of a handgun or stock end of a *long gun* opposite the muzzle end. Compare with *muzzle*.

bystander

an individual in close proximity to a crime or accident who has the potential of witnessing or affecting the response of an incident. Compare with *innocent bystander.*

cache
a store of drugs, weapons, or other illegal goods.

cadaver
a human corpse, especially one to be used for dissection in medical schools.

cadaveric spasm
the physiological reaction that can occur soon after death when a single muscle group becomes stiff and rigid.

Cali Cartel
an infamous illegal drug organization based in Cali, Colombia. At one time, it was thought that the Cali Cartel controlled the vast majority of the world's cocaine distribution. Its influence extended to government officials, including those in the military, the police, and other organizations and individuals. See *cartel, Medellin Cartel*.

caliber
The inside diameter of a rifle or handgun barrel or the diameter of a bullet or projectile. For example, the bore of a .22 rifle is 22/100 of an inch in diameter. Compare with *gauge*.

California Youth Authority
the state agency in the state of California responsible for the confinement of youth remanded by county juvenile courts. The California Youth Authority distinguished itself in the latter part of the 20th century by participating in innovative programming including the use of *I-level*.

call for service
a call received by a law enforcement agency indicating that someone needs police assistance. Calls for service are often employed by researchers and policy analysts as an alternative to crime statistics

such as those gathered for the *Uniform Crime Reports*. Plotting calls for service on a map can aid in the identification of *hot spots*.

call girl

a *prostitute* who uses the telephone to schedule appointments with clients. Call girls generally are accorded higher status than street prostitutes, a difference reflected in their education, income, and lifestyle. Heidi Fleiss, the infamous "Hollywood Madam," operated a lucrative call girl service for show businessmen and other well-heeled customers in Southern California until she was convicted on income tax-related charges.

Cambridge-Somerville Youth Study

a research study of a delinquency prevention project that began in 1937. The Cambridge-Somerville Youth Study was unique at the time because it carefully examined a large number of youth over a long period of time. Data collected on these youth included intelligence, personality characteristics, school progress, and neighborhood characteristics.

cameras in the courtroom

the use of video cameras to record court proceedings, most often trials. Cameras in the courtroom are controversial in that they can disrupt the proceedings and possibly taint the viewing public's perception of what has transpired. Such controversies pit the freedom of the press as expressed in the First Amendment against the right to a fair trial as guaranteed in the Sixth Amendment. See *Court TV.*

Campaign for an Effective Crime Policy (CECP)

an initiative of criminal justice professionals and policymakers to move toward a "less politicized, more informed debate" on criminal justice policy. The CECP questioned unnecessarily long prison sentences, punishment without treatment for substances abusers, the easy access to guns, and the elevation of highly publicized crimes to policy importance.

campus crime

crime occurring in or around a college or university campus. Although campus crime encompasses all such crime, it has come to mean those types of crime common to campus settings, such as *date rape*, bicycle theft, and offenses associated with *binge drinking*.

campus unrest

a term used to convey a variety of disturbances occurring on or around college and university campuses. Examples of campus unrest include demonstrations against war, environmental policy, or other issues. See *civil disobedience*.

caning

the practice of punishing offenders by beating them with split cane shafts. Caning is particularly painful because as the cane is withdrawn, the splits close, pulling flesh with them. The most famous recent case was that of Michael Fay, whose caning in Singapore generated worldwide attention and pleas for leniency. See *flogging*.

cannabis sativa

the scientific term for *marijuana*.

cannibalism

the act of one human being eating the flesh of another human. The most notorious practitioners in recent years were *Jeffrey Dahmer*, a *serial killer* who kept body parts of his victims in his refrigerator, and the fictional Hannibal Lector from the book and movie *The Silence of the Lambs*. In most cultures, cannibalism is considered an offense *malum in se*. See *necrophagia*.

canvass

the process used by law enforcement officers to gather information about a crime or other incident under investigation by asking questions of witnesses, nearby residents, and other persons in close proximity to a *crime scene*.

capacity

the ability of a firearm's magazine to hold a specific number of cartridges. Some jurisdictions limit the capacities of magazines by law in an effort to limit the weapon's lethality. In corrections, the number of detainees or inmates a facility can, should, or does hold. See *design capacity, rated capacity*.

capias

a legal document issued by a judge or magistrate ordering the arrest of an individual.

capital crime

a crime for which the penalty is death. See *capital punishment, death penalty*.

capital offense

same as *capital crime.*

capital punishment

punishment that results in the death of the convicted offender. Capital punishment has become more controversial with exoneration of wrongfully convicted individuals through the use of *DNA testing.* See *death penalty*.

Capone, Al(phonse)

(1899-1947) a notorious *organized crime* figure of the 1920s and 1930s. Capone rose through the ranks of the Chicago mob to become one of the most powerful bosses and bootleggers in the United States. He eventually was convicted of federal *income tax evasion* and was sentenced to prison.

caporegima

literally, the head of a regiment or a lieutenant in an *organized crime* family.

car bomb

an explosive device placed in, under, or near a motor vehicle for the purpose of killing or injuring those in or around it. Car bombs, most closely associated with organized crime and terrorists, can have devastating effects not only to occupants but also to passersby due to the shrapnel created by car parts. The most infamous car bomb, actually transported in a rented truck, was that used in the bombing of the Murrah Federal Building in Oklahoma City, Oklahoma, in 1995. In the United States, car bombs are investigated by the *Bureau of Alcohol, Tobacco and Firearms (BATF).* See *Oklahoma City bombing.*

carbine

a light, short-barreled rifle. An example is the M-1 carbine, a .30-*caliber* version of the longer, heavier Garand rifle used by U.S. forces in World War II.

career criminal

an offender who commits a variety of crimes over an extended period or who specializes in a particular type of crime such as robbery. Career criminals have been the subject of legislation specifically designed to punish them more severely. Also referred to as habitual criminal or *habitual offender*.

carjacking

the illegal seizure of a motor vehicle from its occupants, usually by means of force or threat of force. Carjacking has resulted in the passage of legislation in many states designed to punish carjackers more severely. It has been suggested that carjacking came about as a result of increasingly sophisticated auto theft devices that prevented conventional *auto theft*.

carnal knowledge

to know another in a sexual way.

cartel

an organized criminal enterprise that often specializes in a particular type of crime, such as drug manufacturing and trafficking. See *Cali Cartel, Medellin Cartel*.

cartographic school

an approach to the analysis of crime that makes use of geographical and social data. The cartographic school began in the mid-19th century.

case law

the law that emerges from interpretations by judges when deciding cases. This includes legal principles arising from decisions in appellate courts. As nonstatutory law, case law is also known as common law.

case study

an intensive, empirical examination of a specific set of circumstances to learn detailed information. Case studies yield more indepth information than other research approaches such as surveys.

caseload

the number of offenders a probation, parole, diversion, or other officer supervises or refers for services. These professionals often cite high caseloads as a source of frustration because they inhibit the ability to effectively supervise offenders or render services.

caseload reduction

any program or strategy designed to reduce the number of criminal court cases awaiting disposition, especially those in danger of exceeding time limits. *Alternative dispute resolution (ADR)* is one means used for caseload reduction. Also referred to as docket reduction.

CASKU

see *Child Abduction and Serial Killer Unit*.

Castellammarese Wars

a series of bloody conflicts in the 1920s between Sicilian mafia factions headed by Salvatore Maranzano and Joe Masseria.

castration

the removal or chemical neutralization of male testes employed in the treatment of chronic sex offenders. Physical castration involves the removal of the testes. In *chemical castration*, female hormones such as estrogen are used to compromise the primary sexual characteristics of offenders.

cat burglar

a thief who breaks into and enters a residence or business by stealth, often at night by way of an upper story. See *burglar*.

catharthis hypothesis

the assertion that the violence as reported or portrayed in various media permits the audience to vent their aggressive tendencies and therefore will not have to engage in violent behavior. Compare with *precipitation hypothesis*.

causal fallacy

the fallacy in social policy intended to address crime when the underlying causes are not subject to such policy.

causation

see *etiology*.

cause of death

the injury or disease directly responsible for an individual's death. There can be a primary cause of death, such as a gunshot wound, and a related secondary cause of death, such as massive brain injury.

cautioning

in the United Kingdom, the warning of those who have committed minor offenses. Cautioning is not recommended for those who face serious, indictable offenses. Considerations in cautioning offenders include their age, their view of the *instant offense*, and their *prior record.*

celerity

the swiftness with which punishments for crimes are administered. Celerity is most closely associated with the classical school of criminology. Compare with *certainty, severity.*

cell

the self-contained room in a *lockup, jail, workhouse,* or *prison* in which offenders are housed. Many cells include only a bunk, commode, and wash facilities.

Cellmark Laboratories

a private laboratory located in Maryland that came to prominence in the late 1980s and early 1990s as one of a few agencies and companies capable of analyzing DNA evidence for law enforcement. Cellmark Laboratories has been involved in analyzing evidence in a number of high-profile criminal cases, including the *O. J. Simpson* and *JonBenet Ramsey* cases. See *DNA testing.*

Center for the Study and Prevention of Violence (CSPV)

a center committed to conducting and disseminating research on the causes and prevention of violence. The CSPV, which maintains a literature database that is searchable online, resides at the University of Colorado at Boulder. Its list of Blueprint programs includes those violence prevention initiatives whose effectiveness has been demonstrated through rigorous evaluation.

Center for Substance Abuse Prevention (CSAP)

a division of the U.S. Department of Heath and Human Services responsible for promoting preventive approaches for substance abuse. CSAP makes grants to state and local organizations.

Center for Substance Abuse Treatment (CSAT)

a division of the U.S. Department of Health and Human Services responsible for promoting treatment approaches for substance abuse. CSAT makes grants to state and local organizations.

Centers for Disease Control and Prevention (CDC)

a research and funding arm of the Public Health Service, U.S. Department of Health and Human Services. The CDC, as it is known, oversees an ambitious research agenda on myriad health problems, including homicide, suicide, and intentional injuries. It is also the sponsor of the periodic *Youth Risk Behavior Survey*, a survey of U.S. high school students about their recent experiences with sexual activity and high-risk behaviors.

Central Intelligence Agency (CIA)

the unit of the U.S. government that gathers foreign intelligence information for purposes of national security.

certainty

the aspect of a penal sanction that guarantees its imposition. See *celerity, classical school of criminology, severity*.

chain gang

a form of punishment in which groups of convicted prisoners are forced to perform manual labor while manacled together with iron chains. After having largely fallen into disuse in most jurisdictions, chain gangs saw a return in popularity with the movement toward *retribution* of the 1980s.

chain of custody

the connection of evidence from one handler to another, from the point of collection to analysis and to the subsequent point in the *criminal justice system*. There are a number of ways a chain of custody can be broken. Defense attorneys use broken chains of custody to create reasonable doubt in the minds of jurors as to the integrity of the evidence and the competence of law enforcement authorities.

chalk fairy

a police officer who feels compelled to draw a chalk outline around a homicide victim. Chalk fairies are considered a nuisance by homicide investigators.

chalk lines

see *chalk outline.*

chalk outline

a line drawn by authorized personnel with chalk around a victim at a crime scene to show the location and position of a suspected homicide victim. Accepted law enforcement practice dictates that chalk lines should only be drawn when a body must be moved before the crime scene has been properly photographed, measured, sampled, and documented. See *chalk fairy.*

chamber

the cylindrical compartment of a firearm housing the cartridge when the firearm is discharged.

change of venue

the practice of changing the location of a trial, primarily to avoid prejudicial pretrial publicity that is thought to affect the outcome of the proceedings.

charge

a statement of the offense that an individual is alleged to have committed.

charge bargaining

plea bargaining in which the negotiations center around reducing the type or number of charges or both. It has been suggested that law enforcement officers and prosecutors may overcharge in recognition that the number or severity of charges will eventually be reduced. See *plea bargaining, sentence bargaining.*

check fraud

the deliberate, criminal act of writing checks on nonexistent accounts, closed accounts, accounts other than one's own, or otherwise on an account with insufficient funds to cover the check. See *bad check.* Compare with *passing bad checks.*

check kiting

executing checks without sufficient funds.

chemical castration
the use of Dep-Provera, which contains female hormones, to diminish the sex drive in male sex offenders. Discontinuation of the drug results in the restoration of the offender's normal sex drive.

chemical imbalance
abnormal levels of certain chemicals in the body that can cause behavior changes and various mental disorders such as depression. *Serotonin*, a neurotransmitter, is an example of a chemical that can cause such an imbalance.

Chicago Area Project
a large-scale social experiment designed to address social disorganization and its related symptoms in disadvantaged neighborhoods in Chicago. The Chicago Area Project was grounded in the work of Clifford Shaw and Henry D. McKay. See *Chicago school of criminology, ecological school of criminology, Program on Human Development in Chicago Neighborhoods (PHDCN)*.

Chicago Jury Project
a series of in-depth studies of juries and jurors undertaken at the University of Chicago Law School in the mid-20th century. The Chicago Jury Project shed light on the jury process, whose inner workings previously had been unknown.

Chicago School
a school of criminological thought and associated lines of research that had their beginnings in the Department of Sociology at the University of Chicago. The Chicago School was a product of several prominent intellectuals, including early sociologists Ernest Burgess and Robert Park.

Chicago Seven
seven individuals charged with inciting the disturbances at the 1968 Democratic National Convention in Chicago. The Chicago Seven, defended by William Kunstler, comprised Rennie Davis, David Dellinger, John Froines, Tom Hayden, Abbie Hoffman, Jerry Rubin, and Lee Weiner. Black Panther Bobby Seale was one of the original defendants, but he was tried separately due to his conduct in the courtroom. Several were convicted but the convictions were overturned. See *civil disobedience, riot*.

chief justice
in a federal or state supreme court, the justice who presides over the other justices. In some jurisdictions, the chief justice is elected to the position; in others, he or she is appointed.

Chikatilo, Andrei
a Russian *serial killer* responsible for the deaths of 52 adults and children. The married father of two children, Chikatilo tortured and mutilated his victims. He was executed by a single gunshot in 1994.

child abandonment
the act of a parent deserting or giving up his or her child.

Child Abduction and Serial Killer Unit (CASKU)
a Critical Incident Response Group of the *Federal Bureau of Investigation*. CASKU staff provide investigative support through the development of profiles, investigative strategies, and behavioral assessments as well as offering expert testimony.

child abuse
the intentional harm of a child by a parent or caretaker. Compare with *child neglect*.

child endangering
the intentional or unintentional behavior by a parent, guardian, or caregiver that places a child in danger.

child homicide
a homicide in which the victim is a child. Child homicide differs substantially from adult homicide in that younger children tend to be killed at home by parents or caregivers, whereas older children are more likely to be killed away from home by friends or strangers.

child molestation
illicit, unwarranted contact with a child, usually with sexual intent.

child molester
one who molests children.

child neglect

the failure of a parent to meet the basic needs of a child, such as food, water, clothing, and medical care. Research has shown that child neglect can lead to a higher probability of delinquency and other problems.

child pornography

pornography depicting children. See *Meese Commission*.

child sexual abuse

the sexual exploitation or assault of children by adults.

child witness

a juvenile who is a witness to a crime. Officials in some jurisdictions have made efforts such as videotaping the testimony of child witnesses to spare them the trauma of courtroom testimony.

Children of Murdered Parents

a group of those whose parents were victims of homicide. See *Parents of Murdered Children (POMC)*.

children who witness violence

A term used to describe youth who suffer physical, mental, or emotional trauma as a result of witnessing homicide and other acts of violence. Also, the term is used by programs to identify and treat or conduct research on such behavior. In some cases, a crisis response team, on call around the clock, travels on site to assess such children and refers them for trauma services.

Chinese triads

Chinese secret organizations alleged to control organized crime in New York's Chinatown and other predominantly Chinese communities.

CHINS

Children in need of supervision. CHINS are youthful offenders whose needs are best served by supervision and services rather than by court processing or punishment. See *PINS*.

choke hold

a method of subduing a criminal suspect by tightly seizing the neck, generally from behind. This technique, occasionally used by police officers to restrain a suspect, has been the subject of controversy because of public concern over the excessive use of force by police, especially when the choke hold results in death of the *arrestee*.

chop wounds

wounds in human tissue made by the repeated application of a heavy cutting instrument.

chromatograph

a machine used in *criminalistics* to analyze the composition of chemical mixtures.

chronic offender

a person arrested and convicted of serious crimes so often that he or she is deemed beyond *rehabilitation*. For such offenders, the concern of officials usually is public safety through *incapacitation* rather than rehabilitation, resulting in more severe sentences. Chronic offenders are often subject to longer, more severe prison sentences.

church arson

any attempted or completed burning or bombings of churches, particularly of African American churches by white extremist groups or individuals. Instances of church arson have occurred more often in the southern United States.

circle sentencing

a form of criminal sentencing derived from Native American practices in which the offender permits members of the community to express their feelings about the crime and to make their requirements known. The name derives from the practice of participants arranging themselves in a circle. See *community justice, restorative justice, shaming penalties*.

citizen's arrest

the act of an ordinary citizen placing another person under arrest for a felony. Although citizen arrests are possible and indeed desirable in the case of certain felonies, legal liability accrues to the individual assuming this power.

civil disobedience

a form of nonviolent protest in which a person refuses to obey certain laws or regulations. See *campus unrest.*

civil protection order

a court order designed to protect one person from another. Civil protection orders have primarily been used in cases of domestic violence. Research on civil protection orders suggests that they not only keep most domestic violence victims safe from future abuse but also improve their sense of safety and well-being.

civil remedy

a remedy, other than an indictment in a criminal case, the law awards to a victim against an offender. Examples of civil remedies include financial judgments.

civilian review board

an appointed board whose role is to objectively examine allegations of *excessive force* and other forms of misconduct by police. Civilian review boards have been at the center of much controversy because police often resent the interference from outsiders. See *internal affairs.*

classical school of criminology

a school of criminological thought of the late 18th and early 19th centuries emphasizing free will of those engaging in crime and proportionality between the offense and the punishment. Responding in large part to the arbitrariness and barbarity of the *inquisitorial system*, the classical school favored the abolition of capital punishment and torture. The classical school of criminology, most closely associated with *Cesare Beccaria* and *Jeremy Bentham*, cast the individual as hedonistic, pursuing pleasure and avoiding pain. Humans were considered rational beings, able to consider and choose among various courses of actions. This implied that officials needed only to devise punishments severe enough to deter those contemplating crime. Compare with *neoclassical school of criminology.*

classification instruments

paper-and-pencil instruments designed to assist correctional officials in determining inmate needs and problems and assigning them to appropriate institutions and programs. An example of a classification instrument is the Level of Service Inventory-Revised, which enables correctional officials to assess both risks and needs of offenders.

clemency

an official order by a president or governor setting aside an offender's punishment for a crime. Clemency becomes controversial when those who have committed violent crimes are released, often as a final act of the official before leaving office. President Bill Clinton drew much criticism for granting clemency before the expiration of his second term.

Client-Specific Planning

a process to tailor individualized sentencing and treatment plans for convicted offenders developed by the *National Center on Institutions and Alternatives (NCIA)*. Client-Specific Planning strives to identify a suitable sentencing alternative that incorporates the needs of both the offender and society. It is also used for offenders who are to be released on *parole*. See *privately commissioned presentence investigation reports, rehabilitative ideal*.

Clinard, Marshal B.

a prominent American criminologist of the mid- to late 20th century. Trained at the University of Chicago, Clinard spent most of his academic career at the University of Wisconsin. His contributions include pioneering work on *corporate crime*, cross-cultural study of crime, and, with Richard Quinney, the development of a criminal behavior *typology*.

clinical criminology

the field that applies criminological findings to the assessment and treatment of offenders. In Israel, clinical criminologists are licensed by the ministry of health. See *Client-Specific Planning, privately commissioned presentence investigation reports*.

clinical prediction

the prediction of future dangerous acts or the tendency to offend by use of psychometric tests and other diagnostic tools of the clinician. Clinical prediction has come under criticism for its lack of reliability in determining the likelihood of *recidivism*. Compare with *statistical prediction*.

close security

a level of security in prisons below *maximum security* and above *minimum security*. Prisoners placed in close security typically are escape risks or chronic rule violators. Compare with *maximum security, medium security, minimum security, trusty security*.

Club, The

a steel bar with hooks fastened across the steering wheel of a motor vehicle to prevent its theft. Car thieves have been known to cut through the steering wheel of a car to remove The Club, eliminating its effectiveness. See *crime prevention*.

club drugs

substances of abuse that enjoy widespread popularity among young people who frequent clubs, bars, and all-night dance parties known as raves or trances. Popular club drugs include Ecstasy, Rohypnol, ketamine, *methamphetamine*, and LSD.

Club Fed

informal name for the *Federal Law Enforcement Training Center (FLETC)*.

cluster analysis

a statistical technique designed to identify groups of cases under study. Cluster analysis has been used in criminology and criminal justice to statistically group types of shoplifters, drunk drivers, batterers, and child molesters.

clustered crime scene

a crime scene at which most of the crime-related activity takes place, including the confrontation, the attack, any sexual assault, and the homicide, if applicable.

coagulate

to transform from a liquid to a semisolid. Blood coagulates into a blood clot.

cocaine

a white, crystalline alkaloid, $C_{17}H_{21}NO_4$, derived from cocoa leaves. Although it has medicinal benefits, it is a widely abused drug throughout the world. Colombia is a major supplier, and the United States is a major consumer. Drug cartels in Colombia primarily control the production and distribution of cocaine. See *crack*.

Code of Hammurabi

the earliest written set of laws governing human behavior dating to the 18th century BC. Grounded in Babylonian religious beliefs, the Code of Hammurabi was considered humanitarian for its time. It is said that modern laws derive from the Code of Hammurabi.

coerced confession

a confession extracted from a criminal suspect involuntarily. Coerced confessions were more common before the *Miranda rule*. Such confessions are inadmissible in the United States but may be admissible in other countries, some of which admit the confession as evidence but punish the officer responsible for the coerced confession.

coerced treatment

treatment of substance abuse, mental illness, or other problems that is mandated by judicial or administrative order.

cognitive behavioral therapy

a form of psychodynamic therapy that purports to alter dysfunctional thinking patterns in patients by presenting scenarios. Research has shown that cognitive behavioral therapy is beneficial in correcting thinking errors in offenders.

cohort

a group of people who share something in common such as year of birth. Criminologists study cohorts of people over time to assess their long-term involvement in delinquency and crime. Perhaps the best known such study was that conducted by *Marvin E. Wolfgang*, *Thorsten Sellin*, and Robert Figlio.

Cointelpro

abbreviation for counterintelligence program. A *Federal Bureau of Investigation* initiative to infiltrate, disrupt, and discredit liberal organizations. See *political crime*.

collective conscience

a term used by early sociologist *Emile Durkheim* to describe the beliefs held by a society's citizens. The collective conscience permeates society and is transmitted from generation to generation.

collective efficacy

the tendency of members of a neighborhood or community to look out for one another's interests, including serving as surrogate parents.

collective violence

violent behavior that results from many people coming together, even when many of the participants would not have been violent if acting alone.

collusion

a secret agreement between two or more people for the purpose of fraudulent or deceitful activities.

Columbine massacre

an infamous *mass murder* that took place in April 1999 at Columbine High School in the Denver, Colorado, suburb of Littleton in which 15 people died. Columbine students Eric Harris and Dylan Klebold entered the school armed with semiautomatic weapons, shotguns, and homemade bombs. After fatally shooting 12 of their fellow students and a teacher, as well as wounding numerous others, Harris and Klebold took their own lives. See *school violence, semiautomatic weapons.*

commitment

one of four elements of Travis Hirschi's control theory of delinquency. See *attachment, belief, involvement.*

commodity fraud

fraud involving the trading of goods or merchandise.

common law

the criminal law as interpreted through cases and precedent throughout the ages.

common pleas court

in some states in the United States, courts having original jurisdiction in criminal matters, unless the cases derive from an inferior court such as a county or municipal court.

community aid panel

a group consisting of a *police officer*, a *solicitor*, and responsible community members who assist an offender in addressing the problems that led to his or her illegal behavior. Used in New South Wales, Australia, community aid panels begin their efforts at *rehabilitation* prior to *sentencing*.

community corrections

the spectrum of sentencing alternatives that permit the convicted offender to remain in the community as opposed to serving time in a remote correctional facility. Community corrections include, but are not limited to, community-based correctional facilities, halfway houses, day reporting centers, probation, and parole.

community court

a specialty court in which lay members of the community participate in developing a just disposition for the offender. See *circle sentencing*, *restorative justice*.

community justice

a movement that began in the late 20th century to involve the community in the disposition of criminal cases, based on the premise that the community both gives rise to criminality and suffers the consequences.

community policing

a relatively new philosophy of policing that emphasizes identifying and solving a wide range of community problems that are thought to lead to crime and social disorder. In community-oriented policing, often simply termed community policing, the beat officer and community residents form a bond that permits the regular exchange of information to promote safety and improve the overall quality of life in the neighborhood. Community-oriented policing is closely associated with the late *Robert Trojanowicz* of Michigan State University and with Herman Goldstein of the University of Wisconsin, who coined the term problem-oriented policing. See *problem-oriented policing*.

Community Policing Consortium

a partnership of five police organizations in the United States—the *International Association of Chiefs of Police (IACP)*, the *National Organization of Black Law Enforcement Executives (NOBLE)*, the

National Sheriffs' Association (NSA), the *Police Executive Research Forum (PERF)*, and the *Police Foundation*. The Community Policing Consortium, sponsored and funded by the U.S. Department of Justice, helps advance the philosophy of community policing through research, training, and technical assistance.

community prosecution

the practice of strengthening the relationship between district attorneys and the communities they serve. Starting in the early 1990s, community prosecution programs took the lead of community policing in connecting criminal justice services with neighborhoods.

community service

a sentencing alternative in which the offender performs work for the public good. Community service includes such tasks as picking up trash, painting public buildings, and removing graffiti. Those sentenced to community service generally have to abide by rules such as showing up for work on time, not taking substances of abuse, and performing the required service safely.

community service order

an order issued by a court specifying that an offender participate in *community service*. Community service orders are often part of conditions of *probation*.

community service restitution

see *community service*, *restitution*.

community-based correctional facility (CBCF)

a residential center specifically designed to permit convicted offenders to remain in the community. CBCF residents may receive appropriate drug or other treatment, work regular jobs, and participate in other activities often not part of institutional corrections.

community-based corrections

see *community corrections*.

community-oriented policing

same as *community policing*.

commutation

the process of changing a criminal sentence to make it less severe. A commutation can be granted only by executive authority.

comorbidity

the state of having two or more diseases or conditions at the same time.

compact

an agreement between jurisdictions to accept and supervise convicted offenders. For example, an interstate compact can permit probationers or parolees to move to other states with the understanding that supervision will continue after they move.

comparative criminology

the study of crime and criminal justice processes across other countries and cultures.

comparative justice

the study of criminal justice systems across other countries and cultures.

comparative policing

the study of police and policing across other countries and cultures.

comparison prints

fingerprints taken from possible suspects or others to be compared with *latent prints* found at a *crime scene*. Compare with *elimination prints*.

competency hearing

a court hearing held to establish if a *defendant* is *competent*.

competent

capable of understanding the nature of a crime or crimes for which one has been accused.

complainant

one who swears out a criminal *complaint* against another.

complaint

a formal charge alleging that an individual has committed a particular offense on a particular date.

completed crime

a crime that has been carried out. The *severity* of punishment for a completed crime generally is greater than that for an *attempted crime*.

composite

a picture of a criminal suspect, produced either manually or by computer. For several decades, law enforcement agencies employed artists to render sketches of suspects based on information provided by witnesses. Recently, sophisticated computer programs have permitted police to combine a wide variety of facial features to bring about the same result. Another advantage of the latter method is the ability of nonartistic laypersons to develop composites much more rapidly.

compounding a felony

an agreement to refrain from prosecution in return for financial or other material consideration. Compounding a felony is illegal.

computer crime

a criminal offense facilitated or made possible by the use of a computer. Examples of computer crimes include the electronic transfer of funds. See *computer virus, hacker.*

computer criminal

a person who uses a computer to commit a crime. See *computer virus, hacker.*

computer virus

a computer program designed specifically to annoy the user or cause damage to the computer or its programs or data. Despite the lack of criminal intent in some of these cases, they can be responsible for millions of dollars in damage or other financial losses. Computer viruses have partially been thwarted by antivirus programs that detect and destroy them. The challenge is ongoing, however, since new viruses appear almost daily. See *computer crime, hacker.*

computerized criminal histories (CCHs)

electronically stored and retrievable records of offenders' arrests and case dispositions. CCHs are important for a variety of reasons, including the ability to check the prior records of those applying to purchase firearms.

con

short for *convict*. Also, short for *confidence game*.

con man

short for *confidence man*. One who perpetrates a confidence game.

concealment of birth

hiding the birth of a child. See *neonaticide*.

concealment of death

hiding the death of a person. Deaths are sometimes concealed in order to commit fraud, such as when a family member continues to cash the deceased relative's Social Security or retirement checks.

concentric circle theory

a sociological approach to the analysis of crime, delinquency, and social disorder that lays out cities as a series of concentric rings or circles, with the central city represented by the inner ring. The concentric rings move outward to the suburbs. Each circle represents a different phase of industrial and socioeconomic development. The concentric circle approach was a product of the *Chicago school* and is most closely identified with early University of Chicago sociologists Ernest Burgess and Robert Park.

conciliation

indirect mediation between a victim and an offender. See *victim-offender mediation*.

conditional release

the release of a defendant by the court under certain conditions, such as for participation in treatment programs. See *unconditional release*.

conditions of confinement

the living conditions in a detention or correctional facility.

conduct disorder

a disorder characterized by "a repetitive and persistent pattern of behavior in which the basic rights of others or major age-appropriate societal norms or rules are violated" (American Psychiatric Association, 1994). Conduct disorder is a common diagnosis for a youth who has engaged in violent or other antisocial behavior. Compare with *antisocial personality disorder*.

conduct norms

unwritten rules that guide how people should act. Conduct norms, developed over time through the interaction of group members, vary from group to group. Some conduct norms eventually find expression in laws. Compare with *folkways, laws, mores*.

confession

an acknowledgment of guilt, made either verbally or in writing, by someone accused of a crime.

confidence game

a crime in which the offender elicits the trust of the victim, often playing on sympathy or greed, to relieve the latter of money or other goods.

confidence man

one who operates a *confidence game*.

confidentiality

the principle and practice of not revealing the information shared by one party with another.

confinement

the condition of being locked up in either a short-term or long-term facility for offenders.

conflict model

a theoretical position in criminology that holds that opposing political, social, or other forces in society are responsible for a variety of social ills, including crime and delinquency. Inherent in the conflict model is that there is no normative consensus. Also referred to as *conflict theory*.

conflict of interest
> a situation in which a party to a legal matter or business transaction cannot participate due to a preexisting personal or professional relationship with another party.

conflict theory
> see *conflict model.*

conjugal visit
> an unsupervised visit of a spouse or significant other to a partner who is confined in prison or jail for the purpose of facilitating sexual relations. Conjugal visits for prisoners have been common in Scandinavian countries for many years.

consensus model
> a model in criminology that posits that most people in society subscribe to the same set of norms and values. See *conflict theory.*

consent decree
> a decree issued by a judge that expresses the agreement between two or more parties to a dispute.

consigliere
> in an organized crime family, one who served as adviser to a boss. The role of consigliere was dramatically portrayed by actor Robert Duvall as Tom Hagen in the film *The Godfather.*

conspiracy
> secret, organized effort by two or more individuals to commit a crime. Persons may be charged with conspiracy even though the planned act is never completed.

conspirator
> one who willingly participates in a *conspiracy.*

constable
> a peace officer with limited authority and jurisdiction. In Britain, a police officer.

constitutional theory
> any theoretical perspective in criminology that focuses on the physical, biological, or mental constitution of the person. See *Cesare Lombroso.*

consumer fraud

fraud perpetrated against those who purchase merchandise or services. An example of consumer fraud is shoddy house repair.

contact wound

a wound that occurs when a firearm is discharged while in direct contact with the body. The infusion of gases causes a violent bursting of the skin and a wound with poorly defined edges, defining it as the result of contact. See *starring*.

containment theory

a form of *control theory* advanced by American criminologist *Walter C. Reckless* that posits that individuals are insulated from the risk factors of crime and delinquency by inner containment and outer containment. Inner containment consists of an individual's self-concept and subscription to conventional values and goals. Outer containment consists of the formal and informal strictures placed on the individual by laws and norms. See *good boys, bad boys*.

contamination of evidence

adulteration of evidence by the introduction of a foreign substance, either deliberately or accidentally. Contamination of evidence can result in the withdrawal or dismissal of charges against the accused.

contempt of court

a disregard for the respect and decorum of a court of law. Also, the charge for such disregard. Contempt of court can result in the confinement of the person so found.

content analysis

a set of qualitative data analysis techniques that permits the analyses of text for terms, concepts, themes, or theoretical constructs. Content analysis has been used in criminological research to examine newspaper coverage of certain types of crime.

continuance

a postponement of a hearing, trial, or other court appearance. Continuances may be sought to delay proceedings or simply to annoy the other party.

continuum of care

theoretically, the seamless delivery of services to delinquents or children at risk. A continuum of care improves the coordination and effectiveness of such services.

contrecoup contusion

a contusion of the brain found opposite the point of impact.

control group

in experimental research, the group identical to the experimental group in all respects except for the fact it receives no treatment.

control theory

a set of theoretical perspectives in criminology that assert that certain social forces serve to prevent delinquency and crime by controlling individuals who otherwise would pursue self-interest. Control can be either internal or external. Examples of control theories include *containment theory* and a *general theory of crime*.

controlled substance

any narcotic or other drug whose manufacture, sale, or distribution is regulated by federal law.

contusion

a bruise caused by a blow or impact.

convergence hypothesis

the notion that as males and females move toward equality, the differences between them in the amount of criminal activity will lessen. Thus, as women attain greater sex role equality as measured by such factors as participation in the workforce and education level, they can be expected to engage in more crime, especially *property crime*.

conversion

the act of illegally appropriating the property of another for one's own use.

convict

to prove someone guilty of a crime. Also, a person who has been found guilty of a crime, particularly one who is serving or has served time in prison. See *ex-con*.

convicted innocent

an individual who did not commit the crime for which he or she was convicted. See *wrongful conviction*.

conviction

a finding of a defendant's guilt by a judge or jury.

cooling-off period

the span of time between the offenses of a serial killer. Also, the period between a person's application to purchase a firearm and when he or she actually takes possession, intended to permit any anger on the part of the purchaser to subside.

cop-killer bullets

term given to Teflon-coated bullets capable of penetrating the typical *bullet-proof vest*. Cop-killer bullets have been prohibited by law.

copycat

a person who copies the *modus operandi* of a publicized crime, often a murder, for secondary gain.

copyright infringement

unlawful reproduction or use of copyrighted material.

coroner

a public official charged with the responsibility of investigating all deaths other than those from natural causes. In most jurisdictions, coroners are elected. Their authority to investigate the circumstances surrounding crimes supercedes that of all other local officials. See *medical examiner.*

corporal punishment

punishment that generally consists of a form of hitting or beating. Once common in families and schools, corporal punishment has waned, largely in response to lawsuits by parents and because of the belief that violence begets later violence.

corporate crime

crime perpetrated by a corporation's officials for its benefit. Examples of corporate crime include the production and sale of dangerously defective Pintos by the Ford Motor Company and the dumping of harmful chemicals at Love Canal in New York by Hooker Chemical.

corpse

a dead body, especially of a human being.

corpus delicti

a body that can be seen and possessed. In some jurisdictions, a corpus delicti is necessary in order to pursue homicide charges against a *suspect*.

corrections

the field concerned with the imprisonment, control, and rehabilitation of convicted offenders. Corrections includes the administration and study of prisons, community-based sanctions, parole, probation, and less intrusive alternatives.

corrections officer

one who guards prisoners in a correctional facility. The role of corrections officer has been fraught with controversy and ambivalence: They must face the dangers of guarding violent offenders but are frequently found guilty of engaging in brutality against inmates or smuggling contraband into prison. In most jurisdictions, corrections officers have strong union representation.

corruption

the widespread use of *bribery* or *fraud*.

corruption of a minor

a sex offense involving intercourse or other sexual conduct by an adult against someone under the age of 18. Compare with *statutory rape*.

Cosa Nostra

see *La Cosa Nostra*.

cost overrun

the practice of contractors wherein they knowingly exceed estimated costs for merchandise or services.

counselor

a lawyer who gives advice about the law. Compare with *barrister*, *solicitor*.

counterfeit

illegally produced. For example, counterfeit money is currency that is illegally manufactured.

counterfeiting

the process of producing counterfeit currency or other goods.

county court

a court having jurisdiction over misdemeanors and traffic offenses in a community not having a municipal or mayor's court. County courts generally have jurisdiction over the entire county in which they reside.

court

a government institution whose responsibility is to adjudicate disputes and criminal charges.

court administrator

the person, generally not a lawyer, whose responsibility is to oversee the operation of a court, including budget management, human resources, and docket management.

court backlog

the glut of cases awaiting disposition in a criminal court, some of which may be nearing the legal limits for trial. See *caseload reduction, court delay*.

court clerk

a person responsible for maintaining the records of a criminal court. Court clerks are often elected to office. Also referred to as clerk of court.

court costs

a fee assessed convicted offenders by a judge or magistrate to help recover the costs associated with court processing.

court delay

the delay in processing criminal cases in court. To address court delay, some jurisdictions have adopted time standards for the processing of criminal cases. See *court backlog, speedy trial*.

court liaison service

a program in Australia that provides psychiatric assessment and intervention services to those before the Magistrate's Court. Persons so identified are diverted to the appropriate facility either in prison or in the community.

court reform

a term used to describe any of a wide variety of movements to change the way in which courts are organized or the way in which they operate. Court reform can include such issues as unification.

court reporter

a specially trained stenographer who uses a transcribing machine to record the verbal proceedings in a court of law. In some jurisdictions, court reporters are employed by the court. In others, they are self-employed. Court reporters also transcribe the output from the transcribing machine, thereby making the transcripts available for purposes such as appeals, or for the reading back of testimony to the jury.

Court TV

a television network that debuted in the early 1990s that broadcasts a variety of crime-related programming. Court TV aired the widely viewed flight of *O.J. Simpson* and his subsequent trial. In addition to covering well-publicized criminal cases, Court TV offers other related programming, including reruns of dramatic police and legal series.

court unification

the placement of lower courts under the control of a centralized administrative umbrella. This can include, but not necessarily be limited to, centralization of authority, rule-making powers, budgeting, and funding.

cover-up

The concealment of wrongdoing. Examples of cover-ups include efforts by police to mask the real circumstances of the questionable shooting of a suspect, including the use of a *throwaway* weapon.

CPTED

see *Crime Prevention Through Environmental Design (CPTED)*.

crack

a crystalline form of cocaine smoked in a pipe. Crack was responsible for an unprecedented increase in crime in the 1980s and 1990s. Also referred to as *crack cocaine*.

crack baby

an infant born addicted to crack cocaine as a consequence of its mother's drug abuse. In many jurisdictions, a mother can be charged with a crime simply for taking drugs she should have known would influence or addict her unborn child. Crack babies experience myriad problems, such as low birth weight and withdrawal symptoms if the drug in question is not administered.

crack cocaine

same as *crack*.

crack house

a house or other dwelling, often vacant, used by crack addicts to get high. Crack houses are often the target of law enforcement efforts to clean up socially disorganized neighborhoods.

crank

slang for *methamphetamine*.

Cressey, Donald R.

an American criminologist perhaps best known for studying with and coauthoring *Principles of Criminology* with *Edwin H. Sutherland*. Among Cressey's contributions was a pioneering study of embezzlement and a monograph on organized crime.

crime

behavior prohibited by law and punishable by a term of confinement, the imposition of fines, or other legal sanctions.

crime analysis

the processing of data to shed light on the incidence, prevalence, or seriousness of crime and related behaviors.

Crime Classification System

a typology of violent crime developed by Federal Bureau of Investigation profilers John Douglas and Robert Ressler and nursing professor Ann W. Burgess.

crime clock

heuristic device used by the *Federal Bureau of Investigation* in its *Uniform Crime Reports* to convey approximately how frequently each day each type of serious crime occurs. The use of the crime clock is deceptive because it implies that crime is evenly distributed throughout the 24 hours of a day.

crime displacement

the movement of crime from one area to another due to law enforcement or other intervention efforts. Crime displacement is a consideration in determining the effects and effectiveness of *crime prevention* programs.

crime fiction

fiction whose plot revolves around the perpetration and solution of a crime, most often a homicide. Crime fiction includes several genres of fiction, including *mystery* and *suspense*.

crime forecasting

the practice of using statistical techniques to predict future trends. Accurate crime forecasting can be difficult, however, because future trends do not always mirror past trends. An example of this difficulty was the prediction of a wave of superpredators that would victimize the United States in the early 21st century. So far, no such crime wave has occurred.

crime index

according to the *Uniform Crime Reports*, the total of the most serious crimes, including murder and nonnegligent homicide, rape, robbery, aggravated assault, larceny, motor vehicle theft, and arson. The crime index is used by the *Federal Bureau of Investigation* as a general barometer of how much crime there is in the United States. See *modified index*, *Part I offense*.

crime mapping

the practice of plotting crimes on a map, most often using computer software. Crime mapping permits the visual inspection of crime patterns. See *geographic information system (GIS)*.

crime prevention

the specialized field of study and practice concerned with keeping crime from occurring in the first place. Crime prevention encompasses efforts to reduce opportunities for victimization and those intended to avert youth from delinquency. See *National Crime Prevention Council (NCPC)*.

Crime Prevention Through Environmental Design (CPTED)

Crime Prevention Through Environmental Design is the title of a book by criminologist C. Ray Jeffrey and an area of specialization within the field of crime prevention. The advocates of CPTED maintain that by eliminating the opportunities for crime inherent in the physical environment, many crimes can be prevented. CPTED was originally inspired by the work of noted architect R. Buckminster Fuller. See *crime prevention*.

crime of passion

a violent crime spawned by extreme emotion, such as jealously or rage. See *expressive crime*.

crime rate

the number of crimes per unit of population, most often 100,000. The crime rate per 100,000 is calculated by dividing the actual population of the jurisdiction in question by 100,000. That quotient is then divided into the number of actual crimes. See *Uniform Crime Reports*.

crime scene

the physical area where a crime has taken place. In cases of violent and other serious crimes, great care is usually taken to cordon off the crime scene to preserve evidence that could lead to the identification, arrest, and conviction of the perpetrator.

crime scene cards

cards place at the scene of a crime to announce that the area is restricted to authorized personnel.

crime scene contamination

activities, most often unintentional, that destroy or alter physical evidence and compromise its value for prosecution purposes.

crime wave

a period or trend of a higher amount of crime. Crime waves can be caused by a variety of circumstances, including the introduction of a new drug such as *crack* or the demographic effects of a *baby boom*.

criminal

one who has committed a crime.

criminal alien

an *illegal alien* who has committed or is suspected of having committed a crime.

criminal anthropology

the association of body types and other physical characteristics with the tendency toward engaging in crime. See *Cesare Lombroso, positive school of criminology*.

criminal behavior system

a *typology* of crime associated with criminologists *Marshall Clinard* and Richard Quinney. Their system included the following types of crimes: violent personal, occasional property, occupational, corporate, political, public order, conventional, organized, and professional.

criminal career

the longitudinal sequence of an individual's involvement in criminal behavior. The examination of criminal careers can include such issues as the age at which they begin to offend and the age at which they desist. Also of interest is whether criminals specialize in certain types of crimes or if they participate in a variety of offenses.

criminal investigative analysis

the application of in-depth knowledge about a suspect's personality and behavior to identify and apprehend predatory and other violent offenders. Central to criminal investigative analysis is the development of profiles based on in-depth analysis of the crimes and the offender's behavior. See *Behavioral Science Unit (BSU), Child Abduction and Serial Killer Unit (CASKU)*.

criminal justice funnel

a heuristic device to show the large number of suspected or reported crimes and their attrition as they pass through the various stages of the *criminal justice system*. The large neck of the funnel represents all offenses.

criminal justice system

the entire governmental apparatus that formally processes crime, including but not limited to law enforcement, prosecution, defense, the courts, and corrections.

criminal law

the body of statutes that covers acts defined as criminal.

criminal trespass

a criminal offense, most often a misdemeanor, that involves the uninvited presence of an intruder. Compare with *breaking and entering, burglary*.

criminalist

a scientifically trained person who specializes in the detection or solution of crime through the analysis of physical evidence, such as DNA, fingerprints, tool marks, and body fluids, and other forms of evidence. See *DNA testing, forensic science*.

criminalistics

field concerned with the scientific investigation and detection of crime.

criminality

the quality of being criminal or having such characteristics.

criminaloid

according to the 19th-century Italian criminologist *Raffaele Garofalo*, an offender who is motivated by emotion that, in combination with other factors, results in criminal behavior. See *positive school of criminology*.

criminogenic factors

factors responsible for spawning crime, such as low educational attainment and faulty thinking patterns. Effective correctional programming targets criminogenic factors.

criminological theory

any systematic attempt to explain the causes or *etiology* of crime.

criminologist

a person whose professional identity revolves around the study of crime, criminals, and the criminal justice system. Some people mistakenly call criminologists those who engage in the scientific detection of crime. Compare with *criminalist*.

criminology

the scientific study of crime, criminals, and the criminal justice system. Criminology encompasses not only the *etiology* of crime but also the processing of offenders by law enforcement, prosecution, courts, and corrections.

Crips

a large, Los Angeles-based gang with a reputation for violence. See *Bloods*, *Latin Kings*.

crisis intervention

any of several services intended to assist persons in extreme emotional distress, including those contemplating suicide. Employee assistance programs provide crisis intervention services for employees of organizations.

critical criminology

a criminological perspective that questions the right of those in power to define and control behavior as criminal. Critical criminology was a natural outgrowth of the *labeling perspective*. Compare with *Marxist criminology*.

cross-examination

the examination of a witness by the opposing side, after he or she has initially testified.

cross-sectional study

a research study that takes a one-time snapshot view of the phenomenon of interest. Compare with *longitudinal study*.

cruel and unusual punishment

punishment that exceeds reasonable standards of severity. Cruel and unusual punishment is prohibited by the *Eighth Amendment* of the U.S. Constitution. An example of a punishment that might be considered cruel and unusual is a botched *electrocution* during which the condemned catches on fire and has to have multiple administrations of electrical current before death occurs. Confinement in a *supermax prison* has also been mentioned as cruel and unusual punishment.

cruelty to animals

the gross neglect, abuse, or killing of domesticated animals. One of the three indicators of psychopathy, the other two being *bed-wetting* and *fire setting*.

crystal meth

the crystalline form of the drug *methamphetamine*. Crystal meth gained popularity in the United States in the 1980s and 1990s.

CSA

the Covenant, Sword, and Arm of the Lord, a right-wing extremist group.

culpable

legally blameworthy for a behavior in question.

culture conflict

the termed coined by 20th-century sociologist *Thorsten Sellin* in his monograph, *Culture Conflict and Crime*, to describe the clash of norms in those who emigrated to the United States from abroad. To illustrate his theory, Sellin related the story of the Sicilian immigrant who killed the young man who impregnated his daughter. To the immigrant, he was only doing what any self-respecting Sicilian father would do in the circumstances, but under U.S. norms he was guilty of murder. See *primary conflict, secondary conflict*.

culture of terror

term used to describe the violent world of underground drug trafficking in large cities.

cumulative disadvantage

the net effect of poverty; social disorganization; physical, sexual, or emotional abuse; or other problems associated with inner-city urban life on those who live there.

Customs Service, U.S.

the federal enforcement agency responsible for protecting U.S. borders. The U.S. Customs Service combats *smuggling* using an extensive land, air, and marine force. Due to its broad charge, it investigates narcotics trafficking, prohibited animals and vegetation, pornography, *cybercrime*, and other forms of criminal activity.

cybercrime

crime committed with, or facilitated by, computers, the Internet, the World Wide Web, and other online services. See *computer virus, hacker*.

cycle of violence

the exposure of children to violence by their families that in turn increases the likelihood that they too will engage in violent behavior toward others, including their children, when they grow older. Compare with *intergenerational transmission of crime*.

Dahmer, Jeffrey

(1959-1994) a notorious *serial killer* who, primarily in Milwaukee, Wisconsin, murdered a number of young gay males. Dahmer set himself apart from most other serial killers by torturing, sexually assaulting, and dismembering his victims, cannibalizing and refrigerating some of their remains. Convicted of 16 counts of murder in 1991 and sentenced to life in prison, he was later murdered by another inmate.

daisy chain scam

an illegal operation in which companies create a chain of affiliates, each selling products to another, to manipulate the product's price.

Dalkon Shield

an intrauterine device (IUD) that was widely distributed and marketed by the A. H. Robins Company despite the fact it was responsible for the deaths of a number of women. See *corporate crime*.

dark figure of crime

same as *hidden crime*. See *reported crime, victimization survey*.

date rape

The unlawful sexual violation of an individual during the course of a date. The rapist may use any of a variety of methods to get the victim to submit, including force, threats, or drugs. Compare with *acquaintance rape*.

day fine

a financial punishment in which the offender pays a portion of his or her daily income. The amount paid also depends on such factors as aggravating or mitigating circumstances and the seriousness of the offense. Day fines, which originated in Scandinavian countries, have been shown to not jeopardize public safety.

day reporting center

a correctional program that permits convicted offenders to live at home but receive services at a center. Day reporting centers can take a variety of forms and can be combined with other correctional options, such as electronic monitoring and intensive supervision probation. The primary advantage of day reporting centers is their cost savings over prisons, halfway houses, and other residential correctional options.

day treatment

a rehabilitative alternative in which the convicted offender undergoes correctional treatment usually at a day treatment center during the day, returning to his or her home at night.

dead man walking

an expression that conveys a prison inmate condemned to death on his way to execution. The phrase became popular with the publication of a nonfiction book, *Dead Man Walking*, by Sister Helen Prejean and the subsequent motion picture of the same name.

deadly force

force used by law enforcement officers that can result in death. Concern over the use of deadly force has spawned professional and lay interest in the development of *nonlethal weapons*.

death certificate

an official document issued by a coroner and signed by a physician attesting to the nature and cause of a person's death. Death certificates typically include not only identifying information, such as age, gender, and race, but also the primary and secondary causes of death. See *coroner*.

death penalty

punishment for a crime that results in the execution of the defendant.

death row

the cells where prisoners sentenced to death spend their time awaiting execution. See *dead man walking*.

death row inmate

a convicted offender awaiting *execution* on *death row*.

deathbed confession
revelations of criminal responsibility by one who is about to die. See *dying declaration*.

debtor's prison
a penal facility used in the past to punish those who could not meet their financial obligations.

decarceration
the movement or practice of reducing offender populations in correctional facilities.

deceased
dead. One who is dead.

decomposition
the natural breakdown of organic material upon cessation of life.

decriminalization
the removal or reduction of criminal penalties for acts previously defined in the law as illegal. Perhaps the best known example is the movement to decriminalize *marijuana*. See *National Organization for the Reform of Marijuana Laws (NORML)*.

defeminization
the removal of a woman's breast(s) or other female organs during an assault or homicide. Such behavior, which is typical of *disorganized offenders*, usually occurs postmortem. See *depersonalization, serial murder*.

defendant
a person facing criminal charges.

defense attorney
a lawyer whose responsibility is to ensure that the rights of the accused are upheld. See *appointed counsel, public defender*.

defense witness
a witness whose testimony is believed to serve the interests of a defendant. Compare with *prosecution witness*.

defense wounds

cuts or other injuries sustained by the victim of an attack in attempting to ward off the blows. Defense wounds most often are found on the hands and arms.

defensible space

coined by Oscar Newman in a 1972 book by the same name, defensible space conveys the immediate environment surrounding an individual that should be free of the threat of crime and other risk factors. See *Crime Prevention Through Environmental Design (CPTED)*.

definite sentence

a sentence consisting of a fixed term of confinement. Also referred to as a fixed sentence. Compare with *indefinite sentence*.

deinstitutionalization of status offenders (DSO)

the movement to keep *status offenders* from being held in secure confinement. DSO is predicated on the notion that security is harmful to those who have not engaged in delinquency.

delinquency

acts by a juvenile that, if committed by an adult, would be treated as criminal. Compare with *status offense*.

delinquent

one who engages in *delinquency*.

delinquent subculture

a subset of youth in society who are set apart by a set of beliefs, values, and activities that are often vastly different from those of conventional society. See *subculture theory*.

demand reduction

efforts intended to reduce potential or existing substance abusers' desire to take drugs. Demand reduction depends on educating practicing or potential drug users about the legal, health, and other consequences of drug use. Demand reduction can also encompass treatment, which gained momentum in the early 21st century with California's Proposition 36, legislation mandating treatment over imprisonment for substance abusers. Compare with *supply reduction*.

demography
the field that analyzes the characteristics of human populations.

demonology
the notion that some persons, including those who commit crime, are possessed by evil spirits. At one time, demonology was expressed as secret accusations, torture, and other severe tactics to identify and punish those suspected of being possessed. See *goths*.

dependent variable
in explanatory research, the factor to be explained. Compare with *independent variable*.

depersonalization
efforts made by a killer to cover or otherwise obscure the identity of a victim. See *disorganized offender*.

depraved
mentally sick.

depraved heart murder
extremely negligent or atrocious behavior that results in the death of another. An example of depraved heart murder is a man who wrests a life jacket from a child to save his own life, thereby leaving the child to drown.

depraved indifference murder
see *depraved heart murder*.

deprivation model
a model advanced by sociologist Gresham Sykes to explain how inmates cope with the deprivations inherent in prison life. Compare with *importation model*.

deputy
an officer subordinate to a sheriff.

design capacity
the capacity of a prison or other correctional facility as intended by those who designed it. Design capacity, intended to ensure the safety and security of both inmates and staff, is often exceeded due to *overcrowding*. See *rated capacity*.

designer drugs

trendy substances of abuse, often analogs of a drug such as fentanyl. Popular designer drugs are often responsible for emergency room admissions because users are often unaware of the effects and hazards of the drugs. See *club drugs*.

desistance

the refraining from criminal behavior. Criminologists study desistance to determine why offenders discontinue engaging in illegal behavior. See *criminal careers*.

desserts

what one deserves as the result of committing a crime. See *just desserts*.

detainee

one who has been detained by law enforcement or other authorities.

detective

a law enforcement officer, usually working in plain clothes, who specializes in investigating unsolved felonies, such as homicides, sexual assaults, robbery, and burglary, as well narcotics and *vices*.

detention center

a facility for the temporary or short-term confinement of accused or adjudicated youth.

detention home

same as *detention center*.

determinate sentence

a fixed amount of time a convicted offender must serve in prison. The length of a determinate sentence is prescribed by law and cannot be modified by the sentencing judge or by correctional officials. Same as fixed sentence. Compare with *indeterminate sentence*.

determinism

the theory that human conduct is determined by biological or environmental forces. Implicit in determinism is that offenders are less culpable for their actions. See *biological determinism*.

deterrence

one of the four major justifications for punishment, deterrence conveys the legal threat that criminal sanctions purported pose for prospective offenders. Deterrence is thought by many to dissuade would-be offenders from engaging in criminal behavior. See *general deterrence, specific deterrence.*

detoxification center

a treatment facility designed to permit those who are addicted to alcohol or other drugs to safely withdraw from the substance of abuse. Also referred to as detox center.

deviance

acts or behaviors that deviate from what is considered normal or appropriate.

Devil's Island

a settlement and penal colony in French Guiana. Devil's Island was in use from 1852 to 1946. Due to the unhealthy climate, many prisoners died. Few managed to escape from Devil's Island. See *penal colony.*

differential association

a criminological theory put forth by the late *Edwin H. Sutherland* that posits that persons learn to become criminals through their close, intense association with others disposed to criminal behavior.

differential association reinforcement theory

a restatement of *differential association* in light of behaviorism. Differential association reinforcement theory, developed by Robert L. Burgess and Ronald Akers, emphasizes the role of operant conditioning in the way crime is learned.

digital crime

see *computer crime.*

diminished capacity

a defense used by accused offenders that asserts that they could not control their behavior because of a mental or other condition that diminished their capacity to abstain from crime.

diplomatic immunity

insulation from criminal prosecution extended to foreign members of diplomatic corps. Diplomatic immunity causes controversy when those so immune do not have to face criminal prosecution, even for very serious offenses.

direct examination

The first examination of a witness in a court proceeding. Compare with *cross-examination, indirect examination.*

dirty urine

a urine specimen containing prohibited drugs. Drug treatment programs require clients to submit urine specimens to monitor drug use. A dirty urine test finding can result in loss of privileges, expulsion from the treatment program, or, in the case of those under legal supervision, the revocation of *probation.*

disarmament

the removal of arms from those who pose a threat to the safety and security of a community.

discovery

the legal practice by both parties in a trial of sharing information with the other.

discretion

the ability that inheres in criminal justice roles to make subjective judgments that affect those processed in the criminal justice system. Discretion permits officials to circumvent legislative intent designed to treat offenders more harshly, such as *three strikes and you're out.*

discretionary justice

the ability of law enforcement officers, prosecutors, and other functionaries in the criminal justice system to use their own judgment on how or even whether to process offenders or cases. Although discretion is inherent in much of the criminal justice process and perhaps even desirable, it can lead to disparities in the treatment of similarly situated offenders. Its abuse can result in gross miscarriages of justice.

discretionary release

the ability of correctional officials to release offenders when extraordinary circumstances such as overcrowding occur.

disembowel

the cutting of the abdomen in such as way as to permit the intestines to spill out or be exposed. Also *eviscerate*.

disintegrative shaming

shaming that does not attempt to reintegrate the offender back into society. Disintegrative shaming results in *stigma* for the offender. See *integrative shaming, shaming penalties*.

dismember

to cut the limbs from the torso of a human body.

disorderly conduct

a minor charge, generally a misdemeanor, used by police to charge drunken or otherwise publicly disturbing behavior.

disorganized offender

a *serial killer* who tends to be of lower intelligence, kills spontaneously rather than through careful planning, and engages in depersonalization of his or her victims. Compare with *organized offender*.

disparity

an inconsistency in the way offenders or their criminal cases are treated compared to similar cases. See *sentencing disparity*.

disposition

the conclusion of juvenile or criminal court proceedings, often with an adjudication in an adult case or the imposition of a sentence in a criminal case.

disproportionate minority confinement

the confinement in detention, jail, prison, or other facilities of minorities, particularly juveniles, in percentages out of proportion to their representation in the general population. Beginning in the late 1980s, the *Office of Juvenile Justice and Delinquency Prevention* made disproportionate minority confinement one of its priority areas for research and policy development.

dispute resolution

see *alternative dispute resolution.*

district attorney

same as *prosecuting attorney.*

disturbing the peace

a generally minor offense, most often a misdemeanor, consisting of making too much noise in public.

diversion

the formal routing of offenders away from traditional criminal or juvenile justice processing for a specified period of time and under certain conditions, after which their charges will be dismissed. In theory, diversion minimizes the *stigma* associated with criminal conviction. In practice, the use of diversion sometimes leads to *net widening.*

DNA testing

the use of DNA in criminal cases. DNA, which stands for deoxyribonucleic acid, is unique to the individual, making it the near-perfect means of matching a particular offender to a crime. The analysis of DNA is used both to incriminate and to exculpate those suspected and convicted of crimes.

docket

the totality of cases awaiting processing in a criminal court.

domestic murder

the killing of an intimate, most often a spouse or significant other.

domestic terrorism

terrorist acts committed on domestic soil, often by citizens. Examples of domestic terrorism include the *Oklahoma City bombing* and arson committed by *ecoterrorists.* Compare with *international terrorism.* See *Terry L. Nichols, Timothy McVeigh.*

domestic violence

physical or sexual assault of an intimate, most often a spouse, other relative, or significant other.

domestic violence court

a court that specializes in adjudicating cases of domestic violence. See *specialty court*.

domestic violence shelter

a house or other building maintained for the purpose of providing safe housing for women who have been physically or otherwise abused by a spouse or others in a domestic setting.

double bunking

the practice of housing two jail or prison inmates in a cell intended for one. Also referred to as double celling. See *prison crowding*.

downers

slang term for central nervous system depressants and other drugs known to have a calming or tranquilizing effect.

Dream Team

the group of high-profile criminal defense attorneys that worked to defend O. J. Simpson against the charges that he murdered Nicole Brown Simpson and Ronald Goldman. The Dream Team included at one time or another F. Lee Bailey, Johnnie Cochran, Harvard professor Alan Dershowitz, and Robert Shapiro. Ultimately, they were successful in obtaining an acquittal for their client.

drift

according to David Matza in his book *Delinquency and Drift*, the process by which youths in a delinquency subculture move from conventional activities to illegal activities. See *delinquent subculture*.

drive-by shooting

a shooting from a moving motor vehicle. Drive-by shootings, most closely identified with conflicts between gangs, occasionally result in the injury or death of *innocent bystanders*.

driving under the influence

the act of operating a motor vehicle under the influence of alcohol or other drugs. Also driving while intoxicated. See *Mothers Against Drunk Driving (MADD)*.

driving while black

a derogatory expression used to describe the practice by law enforcement officers of pulling over black motorists when there is no violation of the law. See *racial profiling*.

driving while intoxicated

same as *driving under the influence*.

drowning

death caused by immersion in a liquid, most often water.

drug

An organic or chemical compound, which may or may not have therapeutic effects, used for its mind- or consciousness-altering effects. Illegal drugs include *crack, marijuana*, and *methamphetamine*.

drug abuse

the illegal use of substances of abuse. This is also the term for the offense that consists of engaging in such illegal use. Generally, drug abuse as an offense is a *misdemeanor*, punishable by jail time, a fine, or both.

Drug Abuse Warning Network (DAWN)

a large-scale data collection system sponsored by the *Substance Abuse and Mental Health Services Administration (SAMHSA)*. Using data reported by hospital emergency departments and medical examiners, DAWN collects information on what drugs are being used, which ones are related to drug deaths, and which are currently in vogue among users. See *Drug Use Forecasting (DUF)*.

drug cartel

a criminal organization whose activities include the cultivation, manufacture, distribution, and sale of illegal drugs. See *Cali Cartel, cartel, Medellin cartel*.

drug courier

one who transports narcotics or other drugs, often on his or her person. Also referred to as a mule.

drug court

a *specialty court* designed to meet the treatment needs of those suffering from substance abuse.

drug czar

a chief state or federal governmental official in charge of antidrug policy. For the United States, the drug czar is the director of the *Office of National Drug Control Policy (ONDCP)*.

Drug Enforcement Administration (DEA)

a branch of the U.S. Department of Justice responsible for investigating the violation of federal drug laws.

drug kingpin

the head of a criminal organization whose principal activity is the distribution of illegal drugs.

drug legalization

the movement to making the cultivation, possession, and use of certain drugs legal. See *National Organization for the Reform of Marijuana Laws (NORML)*.

drug trafficking

the selling and distribution of narcotics or other drugs for the purpose of making a profit. Penalties for drug trafficking are much greater than those for possession and use.

Drug Use Forecasting (DUF)

a program designed to collect data on jail inmates regarding the nature and extent of their drug use.

DRUGFIRE program

computer technology that permits law enforcement officials to link firearms evidence in shooting investigations. The DRUGFIRE program was developed by the *Federal Bureau of Investigation (FBI)*.

drunk driving

the act of operating a motor vehicle while under the influence of alcohol. See *driving under the influence, Mothers Against Drunk Driving (MADD)*.

drunkenness

a minor criminal charge against those whose intoxication results in a public disturbance. See *public intoxication.*

DSM-III-R

Diagnostic and Statistical Manual, third edition, revised. See *DSM-IV.*

DSM-IV

Diagnostic and Statistical Manual, fourth edition, published by the American Psychiatric Association. The *DSM-IV* is a compilation of all known mental disorders, with information on their diagnosis and prevalence in the population. Its purpose is to assist clinicians with diagnosis and classification. Although the *DSM-IV* is considered the bible of mental health professionals, it has come under criticism for the questionable validity and reliability of some of its disorders.

dual diagnosis

having two problems in need of treatment at the same time. A mentally ill person with a substance abuse problem is an example of someone with a dual diagnosis.

dualistic fallacy

the mistaken notion that criminal populations under study are distinct from the general population, which is assumed to be composed of noncriminals. The dualistic fallacy is exemplified by the once-held belief that criminals came solely from the lower socioeconomic classes. The existence of *upperworld crime* proves this belief to be false.

due process

the notion in Anglo-American law that those accused of a crime have certain rights to fair and impartial processing, including the presumption of innocence.

dueling

the now outdated practice of settling disputes between two individuals by fighting with guns, swords, knives, or other weapons, often to the death. One of the most famous duels in U.S. history was that between Alexander Hamilton and Aaron Burr, in which the former was mortally wounded. Dueling is virtually obsolete in Western societies, having been banned for nearly two centuries.

DUI

driving under the influence of drugs or alcohol. Same as *OMVI*. See *drunk driving*.

dungeon

an unpleasant, often subterranean chamber used for the confinement, torture, or execution of prisoners.

duress

threat or force used illegally to make someone do something against his or her will. Duress is used to mitigate the seriousness of a crime.

Durham rule

the standard that a person charged with a crime is not responsible if the act was the result of mental disease or mental defect. See *M'Naughten rule*.

Durkheim, Emile

(1858-1917) an early French sociologist who introduced the concept of *anomie*. Durkheim, who analyzed French suicide and other statistics for patterns and regularities, argued that crime serves a positive function in society.

DWI

driving while intoxicated. Same as *DUI*.

dying declaration

an oral or written statement made by one who is about to die that may include accusatory or exculpatory information. Dying declarations generally are regarded as factual because it is thought that one who is about to die has little incentive to lie. See *deathbed confession*.

dysfunctional family

a family characterized by marital discord; physical, sexual, mental, or emotional abuse or problems; neglect; or other problems that prevent the family's normal social functioning.

Eastern State Penitentiary

a prison in Pennsylvania intended to provide both incarceration and labor. Opened in 1929, Eastern State Penitentiary became the model for countless other prisons throughout the United States and the world. It was abandoned in 1971.

echo boom

a population boom created by the offspring of a *baby boom*. Echo booms result in larger numbers of persons of offending ages in the population. Also, like a baby boom, an echo boom strains the resources of the criminal justice system.

ecological fallacy

the mistaken tendency to draw inferences about individuals based on aggregate data and vice versa.

ecological school of criminology

any of several schools of criminological thought that focused on the spatial distribution of social problems such as crime and delinquency. One of the most noted was the *Chicago School* for its early work on *concentric circle theory*.

ecology of crime

The geographical and social distribution of crime, as well as the meaning of such distributions.

economic crime

crime that yields a financial benefit for the perpetrator. Economic crime can be committed by individuals and by organizations.

ecoterrorism

violent and property crimes committed in the name of preserving the environment and other natural resources.

ecoterrorist
a person who engages in *ecoterrorism*.

ectomorph
a body type characterized by thinness or slightness of build. See *William Sheldon*.

elder abuse
the physical, mental, or emotional abuse of an older person, most often a parent or grandparent.

electric chair
a chair designed for the *execution* of condemned prisoners by the application of high-voltage electricity. The electric chair has been known to result in botched executions, prompting critics to label it *cruel and unusual punishment*. Many states that employed the electric chair have changed their means of execution to *lethal injection* or have given the condemned a choice.

electrical equipment conspiracy
a major price-fixing conspiracy in the early 1960s by electrical equipment manufacturers, including General Electric and Westinghouse, resulting in heavy fines and jail terms for some of those involved. This conspiracy was one of the first major cases in which corporations were held criminally responsible for the actions of their executives. See *corporate crime*.

electrocution
a form of *execution* in which the *electric chair* is used.

electronic monitoring
the use of an electronic device that permits authorities to limit and monitor the mobility of an offender placed in home detention. Electronic monitoring is used for both accused and convicted misdemeanants and felons. Such devices permit the control of a large number of offenders by a relatively small number of staff members. The technology has not been perfected, permitting false readings by the electronic equipment. Electronic monitoring is referred to as tagging in the United Kingdom. Also referred to as ELMO.

elimination prints

fingerprints taken from those who may have been in or around a crime scene to distinguish the prints of innocent persons from those of possible criminal suspects.

Elmira Reformatory

a correctional institution, established in 1876, that became the model of reformatories for young offenders. What made Elmira Reformatory distinct from earlier reformatories was its emphasis on education in the trades. With the advent of Elmira came the *indeterminate sentence, parole*, and attempts at *inmate classification*.

embezzlement

the surreptitious theft of money by a person in a position of trust. Examples of such a person are an investment counselor or bank teller.

employee crime

crime committed against a business or other organization by a person in its employ. See *employee theft*.

employee theft

the unauthorized taking of goods or services from a business or organization by those in its employ. Research has shown that much employee theft stems from workers' perception that they need to steal to restore a sense of equity between themselves and their employer. See *employee crime*.

endomorph

a body type characterized by excessive weight in proportion to height. Compare with *ectomorph, mesomorph*. See *William Sheldon*.

entrance wound

a wound in human tissue created when penetrated by a bullet or other projectile. See *bullet track, exit wound*.

entrapment

the practice of law enforcement enticing a person to commit a crime who otherwise would not consider engaging in criminal behavior. Entrapment is controversial.

environmental crime

the destruction or contamination of the environment through neglect or purposeful action. An example of environmental crime is the poor disposal practices that render land unusable. Compare with *ecoterrorism*.

epidemic

the widespread contagion of a disease or other socially harmful phenomenon such as crime in a specific area within a specific period of time. See *tipping point*.

Equity Funding scandal

a corporate swindle in which the Equity Funding created thousands of phony insurance policies. Consequently, both reinsurers and stockholders lost millions of dollars. The scandal resulted in indictments and prison terms.

escalation

an increase in the amount of crime an offender commits. Criminologists study escalation to determine the age or point at which escalation occurs. Compare with *desistance*.

escape

the unauthorized or illegal flight from custody of an accused or convicted offender.

escape artist

one adept at escaping from secure devices, such as handcuffs and straightjackets, or places, such as jails and prisons.

escape from lawful imprisonment

see *escape*.

espionage

the practice of surreptitiously gathering information to which the spy is not entitled. See *industrial espionage, Walker spy ring*.

ethics

the science or philosophy of appropriate human conduct.

ethnic cleansing

the policy of some countries or factions to systematically purge or annihilate members of certain races, ethnic groups, cultures, or national origins.

ethnic succession theory

the theory that the control of organized crime passes from one ethnic group to another.

ethnography

the systematic description of social phenomena as the result of close observation and interaction.

etiology

the study of causation. In criminology, etiology focuses on the causes of crime, including individual, situational, environmental, and societal factors. See *nature-nurture debate*.

eugenic criminology

a theoretical perspective and its related policy that suggests that ridding the human population of certain undesirable characteristics would reduce, if not eliminate, crime. Eugenic criminology found partial expression in the extermination policies of the Nazis in the 1930s and 1940s. See *Holocaust*.

euthanasia

same as *mercy killing*.

event history analysis

a set of statistical methods used to analyze change of a phenomenon from one state to another. Event history analysis permits the analysis of longitudinal data even when all cases under analysis have not experienced the event of interest, such as involvement in delinquency.

evidence

physical material that can be used to establish a case for or against a defendant in a criminal case.

eviscerate

to remove the intestines of a victim. Some serial killers eviscerate their victims, usually postmortem, to depersonalize them. See *depersonalization, disorganized killer*.

exclusionary rule

the legal principle that prohibits the use of evidence illegally obtained by police. In some countries, the questionably obtained evidence is used, but the officer who illegally obtained it can be punished.

ex-con

abbreviation for ex-convict. One who has served time in prison.

exculpatory evidence

evidence that serves to prove that an individual did not commit a crime of which he or she has been accused.

execution

the act of carrying out an official order to put a convicted offender to death. See *capital punishment, death penalty.*

executioner

one who puts condemned prisoners to death.

exemplary project

a project deemed so promising or successful that it should be recognized as worthy of replication by other jurisdictions. The former National Institute of Law Enforcement and Criminal Justice of the 1960s and 1970s designated a number of innovative criminal justice practices as exemplary projects.

exhibitionism

the practice of displaying one's genitals to others, most often for sexual gratification.

exhumation

the process of removing from the place of burial human remains to conduct toxicological tests or other necessary investigative procedures, often to confirm the circumstances surrounding the person's death. Exhumation may involve permission of the deceased's relatives, but in some cases it may be ordered by a court of local jurisdiction.

exit wound

the wound in human tissue made by a bullet or other projectile upon leaving the body. The characteristics of an exit wound often differ from those of its corresponding entrance wound depending on the caliber and type of the projectile. Compare with *entrance wound*.

experimental group

in an experiment, the group to which the treatment is administered. An example of an experimental group in a criminal justice evaluation is one that consists of those who participated in diversion or some other intervention. Compare with *control group*. See *Academy of Experimental Criminology*.

expert testimony

testimony offered for either prosecution or defense by a specialist whose expertise bears on the issue in question. See *Frye rule*.

expert witness

one who provides expert testimony in a court of law. Expert witnesses can testify for either the *prosecution* or the *defense*. See *expert testimony*, *Frye rule*.

explosive residue

remnants of chemicals left behind after an explosion. The analysis of explosive residue can tell officials not only the composition of the explosive device but often also the manufacturer of the elements, which can in turn lead to the identification of those responsible for the bombing.

expressive crime

A crime committed out of emotional distress rather than for personal gain. Compare with *instrumental crime*.

expungement

the destruction or sealing of records of adjudication or conviction. In many jurisdictions, the records of juveniles are automatically expunged after a certain period of time. In adult courts, those with a single conviction may apply for expungement after a specified period of time.

exsanguination

loss or draining of blood.

extended jurisdiction

the policy and practice of maintaining correctional jurisdiction over juveniles after they have reached the age of majority, usually 18 years of age.

extenuating circumstance

a circumstance surrounding the commission of a crime that somehow lessens its seriousness.

extortion

the act of forcing payments by force or threats.

extra Y chromosome

a dated, now disproved theory in criminology that males having an XYY instead of the more usual XY chromosomal configuration are more prone to engage in criminal behavior. Research on the extra Y chromosome has been criticized for relying on confined populations that are not necessarily representative of the general population or of offenders in general.

extradition

the legal transfer of an accused from one jurisdiction to the one in which he or she is wanted. An accused may fight extradiction.

extralegal factors

factors unrelated to a criminal or a criminal case but that may influence the disposition of the case. Examples of extralegal factors include age, race, ethnicity, and socioeconomic status, which have nothing to do with the crime. Criminologists have analyzed the influence of extralegal factors on criminal case dispositions. Also referred to as extralegal variables.

eyewitness

a person who visually witnesses a crime.

eyewitness testimony

testimony offered or given by an *eyewitness*. Eyewitness testimony has been shown through research to often be unreliable, leading to the identification and sometimes the conviction of persons not responsible for the crimes in question.

facial reconstruction

the process of taking a human skull and using clay, hair, or other materials to restore it as closely as possible to a likeness of the deceased for purposes of identification. Computer technology has also been employed to reconstruct human likenesses. See *forensic art*.

failure to appear

the charge leveled against those who do not show up for scheduled court hearings. Failure to appear often results in the issuance of a warrant for the person's arrest.

faith-based programs

crime or delinquency prevention or intervention programs promoted by, or affiliated with, the faith community.

Falcon and Snowman

names given to Christopher Boyce and Andrew Dalton Lee, respectively, two spies who sold U.S. military secrets to the Soviets in the 1970s. See *espionage*.

FALN

a radical Puerto Rican group dedicated to achieving Puerto Rican independence from the United States. Their principal means of terrorism was bombings in both Puerto Rico and the United States.

false advertising

commercial advertising that has the capacity to deceive the buying public. False advertising can also hurt competitors of those who engage in deceptive advertising practices. See *bait and switch*.

false arrest

the erroneous arrest of a person by law enforcement authorities.

false confession

a confession given to law enforcement officials by a person who did not commit the crime in question. Those who offer false confession often desire public attention.

false impersonation

passing one's self off as another person.

false imprisonment

the confinement in prison of a person erroneously believed to be responsible for the commission of a crime. See *convicted innocent, wrongful conviction*.

false positive

a predicted outcome that does not occur. In criminological prediction, an example of a false positive is predicting that an offender will reoffend, but he or she in fact does not. This could lead to penal policies that could result in the confinement of offenders who, in fact, would never reoffend.

false pretenses

fraudulent representations made to obtain money, goods, or services.

family court

a court whose jurisdiction encompasses a wide range of family-related matters, including but not limited to delinquency, dependency, and other juvenile cases.

family group conference

a meeting of the offender, victim, their families, and other supportive people to engage in a dialogue in an effort to bring about restoration and healing. Family group conferencing can be traced to New Zealand and Australia, but it has become popular in the United States, even though it takes many forms. See *restorative justice, shaming penalties*.

fantasy

an often recurring mental imagery conjured up by many sex offenders and serial murderers as a prelude to an offense.

fee splitting

fear of crime
the real or perceived extent to which the general citizenry is concerned about their chances of criminal victimization.

Federal Bureau of Investigation (FBI)
the law enforcement arm of the U.S. Department of Justice. The FBI is responsible for addressing a wide array of violations of federal law. In addition, the FBI offers its investigatory services and crime laboratory to state and local authorities in certain circumstances.

Federal Judicial Center
the research and education center of the federal judicial system in the United States. Established in 1967, the Federal Judicial Center is controlled by a board chaired by the chief justice of the United States. It offers education and training programs for judges, attorneys, and nonjudicial court employees including personnel of clerks and probation offices.

Federal Law Enforcement Training Center (FLETC)
a training facility located in Glynco, Georgia, that offers multifaceted law enforcement training for 74 federal agencies.

federal prison
a correctional facility operated by the federal government for the confinement of offenders found guilty of committing federal crimes. Federal prisons include those at Atlanta and Marion, Illinois. See *Alcatraz*.

Federal Witness Protection Program
a program made available by the federal government in which those who testify in criminal cases and whose lives are likely in danger as a result of that testimony may assume a new identify in an undisclosed location. Use of the Federal Witness Protection Program is most closely associated with government cases against the *Mafia* in which the probability of retaliation against any *prosecution witness* is high.

fee splitting
the illegal practice by physicians of making unnecessary referrals to specialists, splitting the fee in return for the referral. See *white-collar crime*.

feeblemindedness

according to *Charles Goring*, defective mental abilities responsible for criminal behavior.

felicific calculus

same as *hedonistic calculus*.

felony

a serious criminal offense that carries a term of 1 year of imprisonment or longer. Felonies include the most serious crimes, such as murder, rape, robbery, aggravated assault, grand theft, auto theft, and arson. Examples of less serious felonies are corruption of a minor, gross sexual imposition, forgery, embezzlement, passing bad checks, and manslaughter. Because felonies are more serious crimes, they generally result not only in more severe penalties but also in the loss of certain civil rights, including the right to vote and hold public office. Compare with *misdemeanor*.

felony murder doctrine

the doctrine that states that any death resulting from the commission of a felony constitutes murder, even if such death was incidental to the crime or accidental. Those involved in such a felony are thus charged with murder, despite the fact that they did nothing purposeful to bring about the decedent's death.

female crime

crime committed by women and girls.

fence

one who engages in *fencing*.

fencing

the illegal buying of stolen merchandise for the purpose of reselling it. It has been argued that more than merely serving as an outlet for stolen goods, fences actually prompt thieves to steal.

Ferri, Enrico

(1856-1929) an Italian criminologist associated with early positivist criminology. Ferri identified four types of criminals: insane, born, occasional, and those motivated by passion. To Ferri's credit, he suggested that criminal behavior was the result of both individual and environmental factors. See *Cesare Lombroso, positive school of criminology*.

fetal alcohol syndrome
physical and mental abnormalities in a fetus caused by the mother's heavy use of alcohol during pregnancy.

fetish
a sexual preference centered around a body part or practice. Common objects of fetishes include certain types of clothing, fur, feet, high-heeled shoes, and long fingernails.

fetishism
the practice of engaging in a fetish.

feuding
an ongoing, often violent conflict between two factions. One of the most famous examples of feuding in the U.S. was that between the Hatfields and the McCoys, which lasted for many years and resulted in numerous deaths.

filicide
the murder of one's child. See *Susan Smith*. Compare with *fratricide, matricide, parricide, patricide, sororicide*.

financial penalty
a penalty that requires a defendant to pay money. Financial penalties include *fines, day fines*, and *restitution*.

fine
an amount of money levied on a convicted offender in lieu of, or in addition to, another penalty. The range of fines is generally established by statute, giving the court *discretion* in the amount actually levied.

fingerprint
the residual pattern left on surfaces by the application of human hands and fingers. Fingerprints are unique to the individual, permitting their use in connecting suspects to crimes. See *arch, whorl*.

fire setter
one who engages in *fire setting*. See *pyromaniac*.

fire setting

the tendency and attraction toward starting fires. Fire setting is one of the three early indicators of *psychopathy*, the other two being *bed-wetting* and *cruelty to animals*.

firearms examination

the in-depth examination of firearms suspected of being used in crimes.

firearms trafficking

the illegal sale and distribution of firearms.

firing squad

a method of *execution* in which several marksmen shoot the condemned to death. Typically, one member of the firing squad will have a blank round, permitting all members to doubt that they actually caused the death of the condemned. Currently, the firing squad is a form of execution in the states of Utah and Wyoming as well as in a number of foreign countries.

first offender

one who is accused, charged, or convicted of a crime for the first time. First offenders generally get more lenient treatment by authorities. Compare with *chronic offender, repeat offender*.

fix

to use influence to avoid or minimize the legal consequences of criminal, traffic, or other citations or to unlawfully influence the outcome of sporting events or games of chance.

flailing

the practice of using split bamboo stalks to beat convicted offenders. Compare with *caning*.

flash bang

a device that explodes with a loud bang and gives off a brilliant light to create a diversion in support of a *high-risk entry*.

flasher

one who exposes his or her genitals to others, usually for sexual gratification. See *indecent exposure*.

flat sentence

same as *determinate sentence*.

flaying

the practice of skinning. Some sexual murderers have been known to flay their victims.

fleeing

attempting to avoid capture by law enforcement authorities.

flick-knife

British for *switchblade*.

floater

a deceased person floating in water. Gases from putrefaction fill the body cavities, forcing the body to rise to the surface and float. The process can take from 8 to 10 days in warm water or 2 or 3 weeks in cold water.

flogging

a form of corporal punishment in which the convicted offender is subjected to multiple lashes from a whip or flail. See *caning*.

floodgate theory

the notion that the decriminalization of a certain behavior will dramatically increase its adoption. The floodgate theory applies to behaviors such as homosexuality and substance abuse.

folk crime

nonserious offenses that are motivated more by social sentiments than by greed, revenge, or other typical motives for crime. An example of such a crime is the automobile assembler who leaves a beer can inside the door of the car he is helping to build to show dissatisfaction with some aspect of society.

folkways

traditional habits of behavior, grounded primarily in meeting basic needs, as identified by early sociologist *William Graham Sumner*. Compare with *mores, norms, laws*.

forced entry

same as *break-in*.

forcible rape

forcible sexual intercourse with another person against his or her will. Forcible rape is one of the seven *index crimes*. Compare with *statutory rape*.

Ford Pinto case

an infamous case in 1978 in which the Ford Motor Company sold its Pinto model despite evidence from crash tests that showed that rear-end collisions would result in fuel tank explosions. It has been estimated that 500 people died as a result of such collisions. It was determined that Ford proceeded with the manufacture of the car after cost-benefit analysis suggested it would be cheaper to pay for the resulting lawsuits than to make the necessary modifications to the Pinto. See *corporate crime, white-collar crime*.

Foreign Corrupt Practices Act

a 1977 federal law that prohibits the payment of bribes to obtain business. The Foreign Corrupt Practices Act was passed in the wake of a number of prominent bribery cases involving U.S. corporations and foreign governments.

forensic art

art as applied to the investigation of crime and the identification of unidentified subjects, including composite drawings, facial reconstruction, and the use of computer-assisted composite programs.

forensic odontology

the analysis of dental structures in solving crime. Forensic odontologists can compare the teeth of homicide victims to X rays on file. They can also sometimes extract genetic material for *DNA testing*.

forensic pathologist

a medically trained professional who is expert at determining the cause of death.

forensic science

the field concerned with the scientific detection and investigation of crime. Forensic science includes but is not limited to ballistics, criminalistics, forensic odontology, and DNA testing.

forgery

to create or alter money or documents for the purpose of deception or financial gain. See *naive forger, systematic forger*.

fornication

consensual sexual intercourse between two people who are not married to each other.

frame-up

an attempt to make it appear that another is responsible for a crime.

franchise fraud

a scheme in which the defrauded persons are led to believe they can make huge sums of money by purchasing what are really bogus franchises.

Fraternal Order of Police (FOP)

a large union of nonexecutive law enforcement officers. The FOP, with 288,000 members, promotes officer benefits and the protection of law enforcement officers.

fraud

the illegal use of misrepresentations to gain an unfair advantage.

free will

the notion that individuals in society, including criminals, freely choose the course. The notion of free will dates back to the writings of 18th-century jurist *Jeremy Bentham*, who asserted that humankind has the ability to choose a course or action, with the choice generally motivated by hedonism. See *classical school of criminology, hedonistic calculus*.

freebase

cocaine purified by heating with ether for the purpose of inhaling the fumes.

Freedom of Information Act (FOIA)

a U.S. law that requires federal agencies to disclose nonclassified or declassified documents to those who request them. It was through the FOIA that the public learned about former *Federal Bureau of Investigation* director J. Edgar Hoover's extensive files on celebrities and politicians.

Freudian theory

psychoanalytic theory advanced by Sigmund Freud. Freudian theory focuses on conflicts in the individual arising during childhood, some of which are thought to lead to criminal behavior.

frisk

to manually search an arrestee or suspect for weapons, drugs, or other contraband. Frisking is necessary to protect the arresting officer. Officers have to be extremely careful not to get pricked by hypodermic needles, especially because they may carry infectious and even fatal diseases.

frotteurism

the practice of rubbing or touching another without consent for sexual pleasure. An example of frotteurism is the person who, moving about in a crowded place, takes advantage of the close quarters by making physical contact with another.

Frye rule

the legal requirement that scientific evidence introduced into court must have acceptance by the larger scientific community. *Bitemark identification* is an example of evidence that might not stand up to the Frye rule because of its lack of reliability. *DNA testing*, however, serves as an example of a technique that does consistently meet the standard. See *expert testimony*.

functionalism

the sociological notion that deviant behaviors such as crime perform a necessary and useful function in society by reinforcing the norms against such behaviors.

furlough

a leave granted an inmate from a correctional facility or halfway house to pursue employment, education, treatment, or other legitimate purposes.

Gacy, John Wayne

(1942-1994) a prolific serial killer operating in and around Chicago in the 1970s. Gacy, owner of his own construction firm, lured and murdered 33 young men, burying their remains under and around his house. He was convicted of 33 murders and executed in Illinois' electric chair in 1994. While awaiting execution, Gacy, who had played a clown earlier in life, painted a number of self-portraits of himself as a clown.

Gall, Franz Joseph

(1758-1828) a German physiologist who developed cranioscopy, which later evolved into *phrenology*. Gall proposed that the shape of the human skull gave clues to the mental and moral attributes of the individual. See *constitutional theory*.

gallows

the structure, most often constructed of wood, from which or on which a condemned person is executed by *hanging*. See *Tyburn tree*.

Gambino, Carlo

for many years, the boss of New York Italian organized crime. Gambino was succeeded by *John Gotti*.

gambling

the engaging in games of chance.

gang

a group of individuals, often organized along racial or ethnic lines, whose members share values and a sense of identity. See *security threat group*

Gang Resistance Education and Training (G.R.E.A.T.)

a program developed by the *Bureau of Alcohol, Tobacco and Firearms (BATF)* designed to help youth refrain from becoming involved in gangs. G.R.E.A.T. works in schools with law enforcement officers who teach a specific curriculum. Compare with *Drug Abuse Resistance Education (DARE)*.

gaol

a dated English term for a jail or prison.

Garofalo, Raffaele

(1852-1934) an Italian professor of criminal law who argued that criminality has an organic basis and therefore is inherited. See *positive school of criminology*.

gas chamber

a means of legal execution in which the accused is placed in an airtight chamber and then sulfuric acid and cyanide crystals are combined, forming lethal gases that bring about death. See *death penalty, electric chair, firing squad, hanging*.

gate fever

the emotional feeling experienced by prison inmates scheduled for release. Gate fever includes anxiety about where they will live, what they will do to earn a living, and whether they will be able to refrain from engaging in crime.

gauge

the caliber of a shotgun, measured by the number of lead balls the size of the inside diameter of the barrel that would comprise 1 pound. For example, a 12-gauge shotgun is so called because it would require 12 lead balls the size of the barrel's diameter to make 1 pound. See *caliber, shotgun*.

gay bashing

the practice of assaulting gays or lesbians. Perhaps the most notorious case of gay bashing was the senseless murder of Matthew Shepard in Laramie, Wyoming, in 1998. See *hate crime*.

gender-specific services

services for at-risk and delinquent girls that meet needs given their age and development. Gender-specific services address such issues as teenage pregnancy, eating disorders, body self-image, and sexually transmitted diseases.

general deterrence

the characteristic of laws or policies that keep those contemplating crime from doing so. General deterrence is essentially a legal threat. Compare with *specific deterrence*.

general theory of crime

a criminological theory advanced by Michael Gottfredson and Travis Hirschi that posits that offenders are those who are low in self-control and who selfishly pursue their own self-interest. According to this theory, the factor most responsible for this is deficient parenting.

genetic fingerprint

the unique genetic characteristics an individual possesses.

genocide

the systematic killing of a people. An example of attempted genocide is the extermination of the Jews by the Nazis. See *ethnic cleansing, Holocaust*.

geographic profiling

the use of detailed geographic information and specialized analytical tools such as the Rigel geographic profiling system to locate offenders. Geographic profiling can be used in any criminal case in which a suspect's movements can be plotted as points on a map.

get tough policies

criminal justice policies intended to treat offenders more harshly. Get tough policies include those that treat juvenile offenders as adults and those that result in longer prison sentences with diminished chance of parole. See *three strikes and you're out*.

gibbet

in England, an upright post with an extended arm from which executed criminals were displayed to the public. The corpses were sometimes left until they rotted, purportedly to serve as a deterrent to would-be criminals.

global fallacy

the tendency to explain all crime with a specific theory. The global fallacy is common in criminology. An example is arguing that a theory adequately explaining homicide also explains sex offending.

It is unrealistic to think that a single theory could explain all crime, even though there are criminologists who claim they have accomplished this.

gold investment fraud

a scam, often perpetrated during difficult economic times, that lures unsuspecting investors into buying gold, silver, or other precious metals.

good boys, bad boys

terms used to refer to the youth studied by *Walter C. Reckless* and Simon Dinitz in their effort to test Reckless's *containment theory*. Schoolteachers in the study were asked to categorize boys in their classes as either good boys or bad boys.

good cop, bad cop

a tactic employed by law enforcement interrogators to elicit a confession from a criminal suspect. One officer plays a hard-nosed investigator who is out to get the suspect while the other pretends to be the suspect's friend. The ploy sometimes results in a confession by the suspect.

good faith exception

an exception to the *exclusionary rule*.

good time credit

days taken off the sentences of convicted offenders for good behavior. In practice, prison inmates receive good time for the absence of infractions. See *bad time*.

Goring, Charles

(1870-1919) a researcher who compared a number of English convicts to college students, hospital patients, and soldiers in an effort to identify characteristics related to criminality. Although Goring did not find significant differences in the physiques of the two groups, he asserted that the convicts were mental defectives. Goring used subjective impressions to arrive at his conclusions about his subjects' mental ability, which lacked the validity and reliability of standardized tests.

goths

youth who wear black clothing, listen to dark music, and otherwise separate themselves from conventional high school groups and activities. Suspicion was cast on goths in the wake of the *Columbine massacre*.

Gotti, John

an infamous organized crime figure of the late 20th century. Working his way up through the ranks of the mob, Gotti became known as the Dapper Don because of his fashionable clothing and later as the Teflon Don because criminal charges never stuck to him. He was finally convicted and sentenced to life without parole.

graduated sanctions

a series of alternatives for convicted offenders, each of which is slightly more punitive and restrictive than the last. Graduated sanctions, which give officials more options than simply sending an individual to jail or prison, can include such alternatives as *intensive supervision probation*, *house arrest*, and *electronic monitoring*.

graffiti

drawing, writing, or painting on buildings or other public surfaces by vandals or gang members. See *tagging*, *gang*.

graft

bribery or similar practices intended to achieve illegal gain in business or politics.

grand jury

a group of citizens, selected from the list of registered voters, who are brought together to consider evidence against criminal suspects. Grand juries operate in secrecy and issue an indictment when the evidence warrants the further pursuit of criminal charges.

grand larceny

theft of goods or services exceeding a legislatively determined threshold that would constitute a felony.

grand theft

see *grand larceny*.

grand theft auto

grand larceny in which the stolen article is a motor vehicle.

grass-eater

a corrupt police officer who accepts the occasional bribe. Compare with *meat-eater*.

Green River killer

the serial murderer responsible for the deaths of 49 women found in and around the Green River in Washington State from 1982 to 1984. Despite many years of investigation and millions of dollars, the case remains unsolved.

grievous bodily harm

serious physical injury as the result of a criminal act.

grifter

one who conducts a *confidence game*.

grooves

in the rifling of the barrel of a firearm, the recessed, slightly twisted troughs between the lands. See *lands, rifling*.

group conferencing

See *family group conference*.

Guardian Angels

shortened name for The Alliance of Guardian Angels, Inc., a non-profit organization of volunteers whose mission is to fight crime and provide positive role models for youth. The Guardian Angels was founded in 1979 by Curtis Sliwa in New York City. The distinctive red berets and T-shirts are meant to promote deterrence in areas known to have high rates of crime. Members are unarmed but are trained in self-defense and will make citizen's arrests if they see a crime being committed.

Guerry, A. M.

A French statistician who conducted some of the first empirical research in criminology. Guerry analyzed a variety of crime data, searching for regularities and patterns.

guillotine

a beheading machine consisting of a heavy slanted blade that drops between grooved uprights that is used for executing condemned individuals, including convicted offenders. The guillotine, which first appeared in the late 1700s, was used as the form of *capital punishment* in France until the 1970s. During World War II, the Nazis used guillotines to execute thousands of prisoners.

guilt

the state of having committed a criminal offense.

guilty

legally responsible for committing a crime.

guilty but mentally ill

a legal status in which the accused is considered legally responsible for the criminal act for which he or she has been charged but that acknowledges that his or her state of mind is compromised by mental problems. See *M'Naughten rule*.

gun buy-back programs

programs offering an incentive, usually cash, to turn firearms over to authorities. Such programs are based on the premise that fewer available firearms will result in fewer firearms-related deaths and injuries. When such programs are offered, it is not unusual for them to yield so many firearms that officials run out of funds. Research on gun buy-back programs shows that they yield guns that are not typically used in the commission of crimes. Although politically popular, they probably have little or no impact on the violent crime they were intended to prevent. Despite such research, gun buy-back programs continue to proliferate.

gun control

the movement to restrict the purchase, possession, manufacture, distribution, or ownership of firearms by private individuals. See *Handgun Control, Inc.*, *National Rifle Association*.

gun court

a *specialty court* designed to more swiftly process offenders whose crimes involved the possession or use of firearms.

gun lobby

in general, organizations in the United States such as the *National Rifle Association* that work to preserve the *Second Amendment* of the U.S. Constitution. The gun lobby contributes large sums of money to politicians to ensure that the *right to bear arms* will be preserved.

gun meltdown

the practice of taking confiscated firearms and melting them in a foundry to prevent their future recirculation and use.

gunshot

The report given off by a firearm. Also, the result of a firearm discharging.

gunshot wound

Trauma to human tissue caused by the impact of a bullet or other projectile discharged from a firearm.

habeas corpus

literally, to take the body. Habeas corpus expresses the legal princi-ple that states that a person not legally detained or confined must be freed.

habitual offender

an offender who continues to engage in criminal behavior. Some states have enacted laws known as habitual offender statutes that define and are designed to control such offenders. See *chronic offender*.

hacker

one who uses computer skills to gain entry to the computer system of another. In some cases, hackers simply do so as an intellectual challenge. In other cases, the hacker may either take or corrupt information or attempt to electronically transfer funds, in the case of a financial institution.

hair trigger

the trigger of a firearm, most often a pistol or revolver, that requires relatively little pressure to bring about the firearm's discharge.

halfway house

a community-based correctional option in which offenders released from prison make the transition back to community life by living in a house with other offenders. Halfway houses operate on the premise that convicts need a period of supervised treatment on release from prison to facilitate their adjustment on the outside.

hallucinogen

a drug that causes hallucinations. Hallucinogens can be natural, such as the psilocybin mushroom, or synthetic, such as *LSD*.

handcuffs

mechanical device that fastens around the wrists and is designed to restrain the mobility of an offender.

handgun

a small, easily concealed firearm designed to be held and fired with one hand. Because of their concealability and disproportionate use in homicide and other violent crimes, handguns have been at the center of much controversy. See *Handgun Control, Inc.*

Handgun Control, Inc.

an organization founded in 1974 that promotes gun safety legislation. Handgun Control, Inc., gained prominence with the involvement of Sarah Brady, wife of former Reagan White House Press Secretary James Brady, who was seriously wounded during an attempt on the president's life in 1981 by *John Hinckley.*

hanging

a form of execution in which the condemned stands over a trapdoor on a scaffold, with a noose tightened around the neck. When the trapdoor is sprung, the condemned falls, breaking his or her neck.

hard labor

a form of punishment in which the convicted offender must perform vigorous physical work as part of the sentence.

hardened criminal

an offender considered beyond rehabilitation or redemption.

harmful error

an error in handling a criminal case that compromises the constitutional rights of the defendant.

hate crime

a criminal offense motivated by hatred of a specific race, ethnic minority, religion, or sexual orientation. The *Federal Bureau of Investigation (FBI)* only recently began to systematically collect data on hate crimes. In many states, hate crimes are codified as offenses distinguished from the core offenses, such as assault, vandalism, or intimidation. *Gay bashing* is an example of a hate crime.

health care fraud
see *Medicaid fraud.*

hearsay
characterized as unreliable; said of testimony.

hedonism
the pursuit of pleasure and the avoidance of pain. See *classical school of criminology.*

hedonistic calculus
the computations individuals are said to make about both the possible pleasure and pain they will derive by engaging in specific behavior. According to the *classical school of criminology,* such individuals will choose the course of action that maximizes pleasure and minimizes pain. Also referred to as *felicific calculus, hedonic calculus.* See *Jeremy Bentham.*

Hell's Angels
an organized group of motorcycle enthusiasts, headquartered in California, reputedly involved in criminal activity, including protection rackets, drug trafficking, murder for hire, and various property crimes.

hemp
slang for *marijuana.*

heroin
a narcotic derived from opium that is usually injected intravenously, inducing a euphoric state in the user. Heroin is highly addictive. See *methadone.*

hesitation marks
found on some suicide victims, these are shallow, slashing-type wounds that run parallel to a deeper, fatal wound caused by a knife or other sharp object. Such marks are generally indicative of second thoughts or hesitation in the final moments before the fatal injury is inflicted. When present on homicide victims, such marks may represent a form of torturing the victim before murder or possibly an attempt by a perpetrator familiar with the concept of hesitation marks to make a homicide appear to be a suicide.

hidden crime

crime not reported to authorities and therefore not measurable by their crime statistics. Compare with *reported crime*. Also referred to as the *dark figure of crime*.

hierarchical linear modeling (HLM)

a set of statistical techniques that permits the simultaneous analysis of variables representing various levels of analysis. For example, in the analysis of delinquency in a community, HLM is capable of simultaneously examining the effects of individual, family, and community factors.

hierarchy rule

the *Uniform Crime Reports (UCR)* rule that specifies that only the most serious crime in a complex offense is recorded. For example, if in a single incident a suspect commits murder, robbery, and rape, only the murder would be recorded for UCR purposes because it is most serious. This is no longer the case with the *National Incident-Based Reporting System (NIBRS)*, which captures all offenses in a single incident.

high sheriff

same as *sheriff*.

high treason

same as *treason*.

high-risk entry

an entry by law enforcement officers in which there is a high probability of armed resistance by the occupants, which in turn poses a higher risk to the officers. See *flash bang*.

high-speed pursuit

the practice by law enforcement officials of engaging in motor vehicle chases of suspects at high rates of speed to effect the suspect's capture. High-speed pursuits are controversial because they have resulted in the deaths and injuries of officers, suspects, and innocent persons. Specialized training in high-speed pursuit is designed to reduce such deaths and injuries.

hijacking

a crime in which a carrier, most often a motor vehicle, is waylaid, diverted, or otherwise illegally controlled for the purpose of stealing transported goods. See *air piracy*, *piracy*.

Hinckley, John

a young man who unsuccessfully attempted to assassinate President Ronald Reagan in 1980. Hinckley's motives for the *assassination* were tied to an obsession with actress Jody Foster and her role in the movie *Taxi Driver*. Hinckley remains confined in a mental institution.

histology

the study of human tissue. Specialists in histology conduct examinations of cells.

hit

a murder for hire or by order. Hits are normally associated with *organized crime*.

hit man

a person who performs murder for hire or under the orders of a superior in organized crime. See *Murder, Inc.*

HITS

see *Homicide Investigation Tracking System (HITS)*.

hoax

a purposeful deception motivated by either humorous or malicious intent.

Hobbs Act

antiracketeering legislation intended to control interference with interstate commerce.

holding cell

a cell designed to temporarily hold an *arrestee*.

hole

an unpleasant, dungeon-like cell in a prison reserved for recalcitrant convicts. See *solitary confinement*.

Holocaust

the experience of the Jews during the systematic attempt by the Nazis to exterminate them. The Holocaust resulted in the deaths of more than 6 million Jews from 1939 to 1945. See *war crimes*.

home detention

the practice of ordering an offender to remain at his or her place of residence as an alternative to confinement in jail.

homeless persons

persons, most often unemployed men and frequently suffering from mental illness, who do not have stable living arrangements. Homeless persons are frequently perceived as threatening, even though they are responsible for little actual crime. See *public order crime*.

homicide

the taking of human life by another. Although often associated with criminal behavior, homicide also includes the lawful taking of life, such as justifiable killings of criminal suspects by police and instances of self-defense by citizens. See *murder, Supplementary Homicide Reports (SHR)*.

Homicide Investigation Tracking System (HITS)

a program operated by the attorney general of the state of Washington to track and investigate homicides and other violent crime having like methods of operation and other similarities. HITS, which contains data from more than 6,000 homicides and 7,000 sexual assaults, has been a significant contributor to the *Violent Criminal Apprehension Program (VICAP)* database of the *Federal Bureau of Investigation (FBI)*. The HITS staff also consults with other jurisdictions on homicide and rape cases.

homicidomania

the aberrant desire to commit murder.

homosexuality

the practice of sexual relations between persons of the same sex.

hoodlum

a street offender, most often a young one.

hooker

A prostitute. See *call girl*.

hooligan

one who engages in *hooliganism*.

hooliganism

unprovoked violence, often associated with soccer and other European athletic events. Hooliganism at sporting events has resulted in numerous deaths and serious injuries.

hormonal imbalance

a condition in which an excess or insufficient amount of a particular hormone can be responsible for criminal behavior.

hostage

a person who is held until ransom is paid or some other condition of release is met. See *Stockholm syndrome*.

hostage negotiation

the art of working to bring about a peaceful resolution to a hostage situation. Those trained in hostage negotiation respond when a person is barricaded with a hostage or is threatening to commit suicide.

hostile witness

a witness in a criminal case who fails to cooperate with the side that subpoenaed him or her.

hot spots

geographical areas in which there is a high number of reported crimes. Hot spots are used to allocate law enforcement resources.

house arrest

an arrest that specifies that the arrestee must remain at his or her residence under specified conditions for a specified period of time. Sometimes, house arrest is combined with *electronic monitoring*.

human ecology

the study of the relationship between people and the physical space in which they live.

human identification

the art of establishing the identity of persons by reconstructing human remains.

human rights

rights considered so universal that they belong to all persons.

human rights violations

violations such as torture that offend the sensibilities of most civilized societies.

Human Rights Watch

an international organization dedicated to protecting human rights throughout the world. Human Rights Watch, headquartered in New York City, conducts investigations into alleged human rights violations, periodically publishing statistics on executions, torture, and other violations. Ending the sexual abuse of prisoners has been one of its focus issues. It is supported by contributions from other organizations and individuals. See *human rights violations*.

human sacrifice

the practice of ritualistically killing a person, often for religious reasons. The most infamous case of human sacrifice in recent history occurred at the Texas-Mexico border in the 1980s.

hung jury

a jury unable to reach a unanimous verdict in a criminal case.

husband beating

one or more assaults on a husband by his wife. See *domestic violence, spousal abuse, wife beating*.

hypnosis

a sleep-like state in which a person is subject to suggestions from the hypnotist. Law enforcement authorities sometimes use hypnosis to help witnesses to a crime recall important details of the crime.

hypomania

a less severe form of mania.

hypothalamic region

the part of the human brain that controls body temperature, thirst and hunger, and other functions.

ICPSR

Inter-University Consortium for Political and Social Research, a social science data repository located at the University of Michigan. ICPSR has in its holdings a large number of data sets related to crime and criminal justice in its *National Archive of Criminal Justice Data (NACJD)*.

identification parade

British for *lineup*.

Identikit

a kit consisting of myriad facial features on overlapping transparencies used by law enforcement to identify suspects, witnesses, and other unknown persons of interest in criminal, missing persons, and other investigations.

identity theft

the assumption by an offender of another's identity through the use of the victim's personal data, such as Social Security number, credit card number, or other identifying data.

ignition interlock

a device connected to the ignition of a motor vehicle that prohibits an intoxicated driver from operating it. Ignition interlock devices are frequently used for those offenders who already have been convicted of *driving under the influence*.

I-level

A term used to refer to both the Interpersonal Maturity Level and the Integration Level of delinquent youth. See *California Youth Authority*.

illegal alien

one who enters and remains in the country without proper authorization.

imitator

same as *copycat*.

Immigration and Naturalization Service (INS)

a division of the U.S. Department of Justice that conducts inspections and investigations related to immigrants. The INS, whose responsibilities include detaining and removing *illegal aliens*, received much attention for its role in the aftermath of the *Attack on America*. See *Border Patrol, U.S.*

immunity from prosecution

protection against criminal prosecution extended to those willing to testify against accomplices.

impersonating an officer

a criminal offense in which the offender illegally poses as a law enforcement officer.

importation model

the model that suggests that the subculture within a prison stems from both what happens in prison and what inmates bring to the experience. Compare with *deprivation model*.

importuning

soliciting for sexual purposes.

imprisonment

the state of being confined in a prison. Imprisonment is among the more severe sanctions meted out by a *judge* at *sentencing*.

incapacitation

one of the four primary purposes of punishment concerned with restricting the offender's ability to reoffend. See *deterrence, rehabilitation, retribution, selective incapacitation*.

incarceration

confinement in a jail, prison, or other detention or correctional facility.

incest

sexual relations between persons related closely by blood.

incised wound

a wound caused by a cutting instrument.

income tax evasion

the intentional failure to report taxable income at the time of filing federal or other tax returns. Income tax evasion in the United States is a felony. Income tax evasion resulted in the conviction of *Al Capone.*

incompetent

the state of being unable to stand trial in a criminal case due to not knowing right from wrong.

incorrigible

beyond the normal control of parents and caregivers. Incorrigible was used as an official status for youths who did not respond to normal authority. See *status offense, delinquency.*

incriminate

to imply, suggest, or prove that an individual committed a crime. See *self-incrimination.*

inculpatory evidence

evidence that reinforces the guilt of the accused. Compare with *exculpatory evidence.*

indecent

sexually explicit material or actions that offend the sensibilities of the average person.

indecent exposure

a criminal offense in which the offender exposes genitalia or other sexual organs to others. Indecent exposure is generally defined in the law as a *misdemeanor.*

indefinite sentence

same as *indeterminate sentence.*

independent variable

in statistical analysis, any variable or factor that helps explain the phenomenon of interest. For example, if one were analyzing the causes of crime, low socioeconomic status might be an independent variable. See *dependent variable.*

indeterminate sentence

a sentence consisting of a range of months or years, with the actual time served dependent on such factors as the offender's rehabilitation and subtraction of good time. Also referred to as indefinite sentence. Compare with *determinate sentence*.

index crimes

according to the *Federal Bureau of Investigation (FBI)*, the seven most serious crimes are murder and nonnegligent homicide, rape, robbery, aggravated assault, burglary, larceny, and motor vehicle theft. In certain circumstances, arson is included as an index crime. Also referred to as index offenses. See *modified index*, *Part I offenses*, *Part II offenses*, *Uniform Crime Reports (UCR)*.

indict

to formally charge, generally with a *felony*, based on the decision of a *grand jury*.

indictment

a document issued by a grand jury charging a person with a crime. Also called *true bill*. See *bill of information*.

indigent

with respect to accused offenders, those without sufficient funds to retain private counsel. See *legal aid society*, *public defender.*

individual deterrence

same as *specific deterrence*.

industrial espionage

the illegal act of gathering and using information from private corporations by their competitors.

infanticide

the killing of an infant. Infanticide is most often committed by parents or caregivers. Compare with *neonaticide*.

informant

one who supplies information to authorities about alleged crimes, often in return for payment or consideration in his or her own pending legal problems.

information

see *bill of information.*

informed consent

agreement to participate in a research study with full knowledge of its purposes and possible physical and psychological consequences.

inhalant

a substance inhaled for its mind-altering effects.

initial appearance

in adult courts, the first appearance of an offender before a judge or magistrate. At initial appearance, the accused is apprised of charges and a bond is set.

injunction

a judicial order prohibiting a behavior or compelling one.

inmate classification

the practice of assigning prison inmates based on such factors as perceived risk.

inmate suicide

the taking of one's own life by an inmate of a jail, prison, or other detention or correctional facility.

innocent

without guilt. See *not guilty.*

innocent bystander

a bystander who has nothing to do with a crime but who gets injured or killed.

Innovative Neighborhood Oriented Policing (INOP)

a program designed by the *Bureau of Justice Assistance (BJA)* to further community-based *demand reduction.* INOP, which was piloted in eight U.S. cities, was grounded in the importance of community, including *community policing* practices.

inquisitorial justice

justice of the past in which secret accusations and torture were employed to identify and convict those suspected of wrongdoing.

insanity

the state of being of unsound mind. Insanity in a criminal justice sense relates to the defendant's ability to distinguish right from wrong.

insanity defense

a strategy employed by a defendant or defense counsel or both in which they maintain the defendant was mentally ill at the time of the alleged offense and consequently did not know right from wrong.

insider trading

the illegal use of confidential stock market information to one's own advantage. Notable cases of insider trading include those involving *Ivan Boesky* and *Michael Milken*.

instant offense

the current offense with which a defendant is charged.

Institute for Law and Justice (ILJ)

the Institute for Law and Justice, a private research and policy organization that undertakes criminal justice studies, performs policy analyses, and consults on various matters related to criminal sentencing and so on.

Institute for Youth Development (IYD)

a nonprofit organization that promotes the avoidance of alcohol, drugs, sex, tobacco, and violence. Central to IYD's philosophy is that youth are capable of making appropriate choices, especially if supported by their families. The IYD is headquartered in Washington, D.C.

institutionalization

the process that occurs in people during prolonged periods of confinement, resulting in their inability to cope with mainstream life outside the institution and, sometimes, the desire to do so.

instrumental crime

a crime such as robbery or theft that yields a demonstrable benefit for the perpetrator. Compare with *expressive crime*.

instrumental Marxism

the Marxist notion that the crime problem stems more from the process of criminalization than from individual or environmental factors emphasized in traditional Western theories.

insurance fraud

a scheme in which a bogus claim is made to an insurance company based on a contrived loss. Insurance companies estimate their losses to insurance fraud in the billions of dollars annually.

intake

the point at which a youth formally becomes involved in the juvenile justice process.

integrative shaming

see *disintegrative shaming, shaming penalties*.

integrative theory

a criminological theory that seeks to combine and reconcile two or more existing theoretical perspectives into a more comprehensive, unified perspective.

intensive supervision probation (ISP)

a form of probation that involves an extra measure of supervision and control by the probation officer. ISP is often used for chronic and other high-risk offenders who pose a greater probability of reoffending. Also referred to as intensive probation supervision (IPS).

intent

the decision to undertake a certain course of action.

intentional injury

a physical injury caused through purposeful action by another. Assault is an example of an intentional injury. The *Centers for Disease Control and Prevention* analyze intentional injuries to determine the causes and correlates of violence.

interdiction

efforts to stem the sale or distribution of illegal goods, most often narcotics and other drugs. Interdiction includes such activities as intercepting motor vehicles, watercraft, and aircraft that are transporting illegal drugs. See *supply reduction.*

intermediate sanction

a punishment for a crime between a mild sanction and a severe sanction. Examples of intermediate sanctions include *electronic monitoring* and *intensive supervision probation (ISP)*, whereas the more mild and severe extremes might be *nonreporting probation* and confinement in *prison*, respectively.

internal affairs

a unit within law enforcement agencies responsible for investigating alleged misconduct and corruption. Internal affairs units have the reputation, deserved or not, of aggressively pursuing "dirty cops," often using questionable methods to achieve their ends.

Internal Revenue Service (IRS)

the federal agency, part of the U.S. Department of the Treasury, responsible for enforcing provisions of the Internal Revenue Code.

International Association of Chiefs of Police (IACP)

an organization of police executives from throughout the world. The IACP sponsors research, training, and holds annual conferences for its members.

international terrorism

terrorism that transcends the boundaries of any one country.

Interpol

International Criminal Police Organization. A multinational collaborative of law enforcement agencies that cooperates in the investigation and apprehension of wanted criminals. Interpol is headquartered in Paris.

interrogation

the informal or formal questioning of suspects or witnesses to a crime.

interstate compact

a written agreement among states and territories.

Inter-University Consortium for Political and Social Research (ICPSR)

a membership consortium of colleges and universities established in 1962 and headquartered at the University of Michigan at Ann Arbor. The ICPSR maintains an extensive archive of social science data sets. It also offers training in quantitative methods and technical assistance to researchers using ICPSR data.

intimate violence

violence between those living together who may or may not be related. See *domestic violence*.

intruder

one who purposefully breaks and enters a premises.

inventory shrinkage

the loss of a company's inventory due to *employee theft*.

investigation

a formal process in which information about an alleged crime is collected and studied.

investigator

a person who conducts an investigation. See *detective*.

involuntary manslaughter

the killing of another through recklessness while committing an illegal act. An example of involuntary manslaughter is accidentally shooting someone to death while *poaching*. Also referred to as negligent manslaughter. Compare with *voluntary manslaughter*.

involvement

one of four elements of Travis Hirschi's control theory of delinquency. See *attachment, belief, commitment*.

Iran-Contra case

a scandal during President Reagan's administration in which the U.S. government sold arms to Iran in exchange for the safe return of hostages. The money received from Iran was then used to fund the Nicaraguan contras. Both the sale of the arms and the giving of money to the contras were against the law. Also referred to as Irangate.

irresistible impulse

phrase used to describe the alleged inability of an individual to control the desire to engage in criminal behavior.

Jack the Ripper

the name of the serial murderer of several prostitutes in London's White Chapel district in 1888. The Jack the Ripper murders were never solved, despite numerous theories on possible suspects.

jail

a short-term facility for holding accused or sentenced offenders. Compare with *lockup, workhouse, prison*.

jailbreak

an escape from a jail or similar holding facility.

john

a term for a man who solicits sexual services from a prostitute.

John Howard

A famous penal reformer of the 18th century.

John Howard Society

a society named for penal reformer *John Howard* dedicated to the humane treatment of convicted offenders. Through its various chapters, the John Howard Society strives for penal reform and promotes programs for offender rehabilitation. It grounds its efforts in six core values that emphasize the redeemability of the offender and the responsibility of all to facilitate their rehabilitation.

Joliet

the site of the famous Stateville Correctional Center in Illinois. Infamous inmates confined at Joliet include *John Wayne Gacy* and *Richard Speck*.

joyriding

the act, often by juveniles, of taking and driving a motor vehicle without permission. Joyriding most often differs from *auto theft* in that those doing it generally intend to return the vehicle after they have had their fun.

judge

a governmental officer appointed or elected to decide questions of law.

judicial discretion

the inherent ability of a judge to make subjective decisions about cases before the court.

judicial misconduct

misconduct perpetrated by judges in the official performance of their judicial duties. One of the best known cases of judicial misconduct was that uncovered by the *Federal Bureau of Investigation (FBI)* in *Operation Greylord* in Cook County, Illinois.

Jukes

a family used in attempts to illustrate the inheritability of criminal tendencies. See *Kallikaks*.

JUMP

see *Juvenile Mentoring Program*.

jurisdiction

an area with geographical boundaries in which law enforcement, prosecutors, courts, or other criminal justice functionaries have authority.

juror

a citizen selected, usually from the list of registered voters, to decide the guilt of a defendant in a criminal case based on evidence presented.

jury

a body of citizens, usually selected from the pool of registered voters, charged with making determinations of guilt or innocence in a criminal trial based on evidence and testimony presented.

jury deliberations

the discussions by jurors in a criminal case that eventually lead to a verdict or to a *hung jury*.

jury duty

the practice and obligation of citizens serving on juries for local courts.

jury instructions

verbal instructions given by a trial judge to members of a *petit jury* prior to commencement of the trial. See *jury nullification*.

jury nullification

the disregard by a jury of the evidence presented and the rendering of its verdict based on other criteria.

jury selection

the process of picking by elimination jurors for a trial by the prosecutor and defense counsel. In highly publicized cases, defense counsel may use the services of a jury consultant. See *voir dire*.

jury system

a system of justice employing a body composed of the accused's peers, who allegedly can hear the facts and render an impartial verdict.

jury tampering

a criminal offense revolving around an attempt to threaten, bribe, or otherwise influence one or more members of an impaneled jury. Jury tampering is most closely associated with organized crime, which has a history of trying to influence jurors to control the outcome of a criminal trial.

jury trial

a criminal trial making use of a *jury* to hear facts and decide the guilt of the *defendant*. Compare with *bench trial*.

just deserts

the principle that asserts that the punishment for a crime should be commensurate with the seriousness of the crime.

Just Say No

an antidrug campaign of the 1980s that rested on the premise that youth and other potential drug users would refrain simply by the repetition of this phrase. Also, the slogan of this campaign. Then First Lady Nancy Reagan was one of the campaign's most vocal proponents.

justice model

a philosophy and its associated policy that emerged in the wake of the 1970s *nothing works* movement. The justice model emphasized predictability of legal consequences for offenders.

justice of the peace

a local official who has the authority to issue licenses, perform marriages, and adjudicate traffic and other minor offenses.

Justice Research and Statistics Association (JRSA)

a national organization whose membership primarily consists of statistical analysis center (SAC) directors. JRSA provides specialized training in criminal justice data analysis techniques, holds an annual meeting each fall, and publishes the journal *Justice Research and Policy*. JRSA's offices are in Washington, D.C.

justifiable homicide

the taking of a life that is excusable due to self-defense or some other consideration.

juvenile

one who is under the age of consent, most often 18 years of age. Compare with *adult*.

juvenile bindover

the transfer of juvenile cases to adult court. Juvenile bindover is used in serious offenses, such as homicide, rape, robbery, and serious assault. Once bound over, the juvenile's case is handled like any other felony case.

juvenile court

a court of law, sometimes a subdivision of a common pleas court or domestic relations court, that has jurisdiction over matters pertaining to the delinquency and unruliness of persons who have not yet attained adulthood.

juvenile delinquency
criminal offenses committed by those under the age of legal responsibility, most often 18 years of age.

juvenile delinquent
term describing a youth who has engaged in *juvenile delinquency*.

juvenile detention center
a temporary holding facility for youthful offenders. Also referred to as an attention center. Same as *detention center*.

juvenile gang
a gang whose members are predominantly juveniles.

juvenile justice
the entire system designed to process and meet the needs of offenders younger than the age of 18.

Juvenile Mentoring Program (JUMP)
a grant program of the *Office of Juvenile Justice and Delinquency Prevention (OJJDP)* designed to promote mentoring to specifically address poor school performance and dropping out of school.

juvenile offender
An offender younger than the age of 18.

juvenile sex offender
a youth who has committed one or more sex offenses. Juvenile sex offenders require different treatment.

Kaczynski, Theodore

a former university professor who was convicted of several fatal bombings in the United States. Kaczynski was a brilliant mathematician who had studied at Harvard University and once held an academic post at the University of California, Berkeley. He was known as the *Unabomber*.

Kallikaks

a family, chronicled by Henry Goddard, that spawned a number of feebleminded and criminal offspring over the course of several generations. Based on his research, Goddard concluded that crime is inherited. See *Jukes*.

kangaroo court

an unofficial court, often held by a mob or vigilante group, whose judgment is biased against the accused.

Kansas City Preventive Patrol Experiment

a progressive law enforcement experiment of the 1970s in which police were assigned to patrol certain high-crime areas in an attempt to reduce crime.

Kefauver Commission

a commission convened by Senator Estes Kefauver in the 1950s to investigate organized crime in the United States.

Kerner Commission

abbreviated name for the National Commission on Civil Disorders. The Kerner Commission, chaired by Senator Otto Kerner, was a federal response to the racial unrest of the 1960s. The Kerner Commission issued its report in 1968.

Kevlar

a synthetic material from which body armor is made. Despite its protective attributes, Kevlar can be penetrated by Teflon-coated bullets or those designed to pierce armor plate. See *body armor, bullet-proof vest*.

Kevorkian, Jack

a Michigan pathologist who gained notoriety by assisting elderly and terminally ill patients to take their own lives. Kevorkian was tried multiple times for murder, eventually being convicted. See *assisted suicide*.

kickback

a payment to an official by a contractor for favoritism in awarding a contract.

kidnapping

the willful deprivation of another's liberty, often to secure ransom. Perhaps the most famous kidnapping of the 20th century was that of Charles Lindbergh's baby.

kill

to bring about the cessation of life. See *murder*.

King, Rodney

a Los Angeles motorist subjected to a videotaped beating at the hands of Los Angeles police officers in 1991. King was stopped for a routine traffic arrest, after which he was beaten by several officers. Several of the officers were later found guilty of excessive force, resulting in suspensions, firings, and criminal convictions. The case led to the formation of a commission to investigate the Los Angeles Police Department.

kingpin

the head of a narcotics trafficking organization. Also referred to as *drug kingpin*.

kleptomania

a psychological compulsion to steal.

Knapp Commission

the commission convened in the early 1970s to investigate widespread corruption within the New York City Police Department. Officer *Frank Serpico* was instrumental in bringing this corruption to public light, resulting in numerous indictments and convictions of police officers. See *police corruption.*

Ku Klux Klan

A racist, white supremacist organization, found mostly in the southern United States, dedicated to asserting the superiority of whites over racial, ethnic, religious, and other minorities, including blacks, Jews, and gays and lesbians.

La Cosa Nostra

another term for the *Mafia*. Joseph Valachi, in his testimony before the *McClelland Commission* hearings in the 1960s, reportedly was the first to use this term to describe the Italian underworld. See *organized crime*.

labeling

the processing of offenders and other deviants so as to attach a new identity that may in turn have negative consequences.

labeling perspective

a theoretical perspective in the sociology of deviant behavior that holds that agents of social control often unnecessarily label those whom they process, sometimes leading to further deviance. Also referred to as labeling theory. See *primary deviance, secondary deviance, stigma*.

labor camp

a camp in which sentenced offenders are required to perform physical labor.

lands

the ridges between the rifling grooves in the barrel of a firearm. Lands, like grooves, leave distinct marks on bullets passing through the barrels of rifled firearms, making possible matches between bullet and firearm. See *grooves, ballistics*.

Lansky, Meyer

a major figure in organized crime in the mid-20th century.

larceny

same as *theft*.

latent function

unintended or unanticipated consequences of laws or rules. A latent function of the law prohibiting the manufacture and distribution of alcoholic beverages might be the development of a *black market* for alcohol. Compare with *manifest function*.

latent prints

fingerprints lifted from surfaces other than the hands bearing the prints. See *arch, fingerprint, whorl*.

Latin Kings

a notorious *gang*. See *Bloods, Crips*.

Law Enforcement Assistance Administration (LEAA)

a large agency in the U.S. Department of Justice created by the *Omnibus Crime Control and Safe Streets Act of 1968*. The LEAA distributed millions of dollars for various crime control programs, many of which drew criticism for their waste of funds on such purchases as armored personnel carriers for small-town police departments.

law-related education

a large-scale, coordinated effort to lessen youth violence and delinquency by involving youth in learning about the law and developing alternatives to violence, substance abuse, and other maladaptive behavior. See *Youth for Justice*.

learning theory

any one of the several theories in criminology that assert that offenders learn criminal behavior in association with other offenders by principles of operant conditioning or other methods.

leg irons

manacles designed for use on the legs. Leg irons, especially those used in the past, were heavy and constrained the person's ability to run.

legal aid society

a private, nonprofit organization of attorneys that exists to represent the rights of indigent clients. See *appointed counsel, public defender*.

legal dose

the amount of drug recommended to treat the condition in question.

legal factors

also legal variables. See *extralegal factors*.

legalization

the process of making legal that which previously was illegal. Compare with *decriminalization*.

Lepine, Marc

a Canadian mass murderer who shot and killed 14 female students at the University of Montreal in 1989. See *mass murder*.

lethal force

force used by law enforcement, prison guards, and other criminal justice officials that could result in the death of suspects or inmates.

lethal gas

the combination of cyanide and sulfuric acid used to execute condemned prisoners. See *gas chamber*.

lethal injection

a form of capital punishment in which lethal amounts of prescription drugs are administered to the condemned until life ceases.

lex talionis

the principle in criminal law that embodies the notion of retaliation. Lex talionis implies that society's response to criminal offenses will be commensurate with the seriousness of the offense and the harm done.

libel

a false, published statement that damages the name or reputation of another.

lie detector

a machine that purportedly reveals deception in individuals by measuring and recording changes in blood pressure, pulse rate, and respiration. The use of lie detectors is controversial because of their unreliability in uncovering the truth. See *polygraph test*.

life imprisonment
same as *life sentence.*

life sentence
a judicially imposed term of confinement equal to the life of the sentenced offender. Offenders who receive a life sentence will die in prison unless their sentences are commuted or they are granted a *pardon.* See *commutation.*

life without parole
a term of confinement in a penal institution that does not include the possibility of release by parole authorities. Life without parole practically means that the offender so sentenced will die in prison.

lifer
an inmate who has been sentenced to life imprisonment.

lineup
an array of persons for the purposes of suspect identification. A lineup consists of not only the suspect but also other individuals, sometimes police officers in street clothes. The victim or witness tries to identify the suspect from this array. See *eyewitness identification.*

linkage blindness
the inability of law enforcement officials to make connections between related crimes. Linkage blindness is a problem in serial crimes, such as homicide or sex offenses, where offenses occur in multiple jurisdictions. Greater participation in the *VICAP* (Violent Criminal Apprehension Program) might help reduce or eliminate linkage blindness.

littering
the offense, most often a misdemeanor, of discarding trash or other material in public.

LiveScan
a patented, electronic system for scanning human fingerprints. Based on complex mathematical algorithms, LiveScan captures the precise contours of the fingerprint, preserving it for later comparison with other prints.

lividity

The collection of blood in a corpse caused by gravity. The presence of lividity can be used to determine if a body has been moved after the blood has settled. Also *postmortem lividity*.

loan sharking

the lending of money at rates of interest far above what is considered reasonable. Generally, those who borrow in these circumstances have a very limited time to pay both the principal and the interest. Failure to pay can result in serious injury or even death.

lobotomy

a psychosurgical procedure in which a cut is made into the frontal lobe of the brain, rendering the patient less violent and more docile. Once common in psychiatric treatment, the procedure has fallen into disuse. See *psychosurgery*.

lockdown

a procedure in a prison in which officials lock all prisoners in their cells. Prison officials use lockdowns when they have reason to believe inmates possess weapons or other contraband or that inmates are planning to create a disturbance.

lockup

a holding position or facility for offenders, often found within a prison, often designed for short-term stay. See *jail, workhouse*.

loitering

a petty offense characterized by an individual's presence in a public or private place without official business or other reasonable justification.

Lombroso, Cesare

(1835-1909) an Italian physician and criminologist of the 19th century who asserted that criminals were throwbacks to an earlier form of human. Lombroso's contributions to the study of crime included an emphasis on the systematic collection of data.

long gun

a firearm with a stock and long barrel, such as a rifle or shotgun. Compare with *handgun*.

longitudinal study

a study of youthful or other offenders in which personal and legal data are periodically collected at various points during the life course. Longitudinal studies are expensive and take a long time to yield definitive results. Other problems with this approach include trying to stay in touch with subjects who periodically move and whose memories may not serve them well when asked by researchers to recall certain events. Notable examples of longitudinal studies are the National Youth Study administered by the University of Colorado, the Pittsburgh Youth Study administered by the University of Pittsburgh, and the Rochester Youth Study administered by the State University of New York at Albany.

lookout

a person who keeps watch as other offenders engage in criminal conduct.

loop

a part of a human fingerprint. See *arch, whorl.*

Low Jack

a brand name of an electronic device installed in a motor vehicle that enables authorities to locate and track it if it is later lost or stolen. Despite their role in recovering stolen vehicles, Low Jacks and similar devices are controversial because of the "big brother" aspect of knowing the motorist's whereabouts at any given time.

LSD

abbreviation for lysergic acid diethylamide. Commonly called acid. A hallucinogenic drug highly favored by the youthful counterculture in the 1960s, it was popularized in song by many psychedelic recording artists and was the drug of choice of drug guru and pop icon Timothy Leary.

lynching

the practice of illegally taking the life of another by hanging, generally accomplished by a mob and often motivated by racial or ethnic hatred. Now infrequent, lynching in the United States is associated with white supremacists and their targeting of blacks. See *hate crime, Ku Klux Klan.*

mace

an aerosol chemical designed to render an attacker temporarily immobile and harmless. See *pepper spray*.

MADD

Mothers Against Drunk Driving, a national organization dedicated to fighting drunk driving through publicity campaigns and the passage of legislation. Many members of MADD have lost loved ones in drunk driving accidents. In fact, MADD is composed of not only mothers but also fathers and other individuals concerned with stopping drunk driving and underage drinking.

Mafia

organized crime of Italian or Sicilian origin.

magazine

the often detachable container that holds cartridges for a firearm. The capacity of magazines has been the subject of debate and legislation, the rationale being that higher-capacity magazines increase *lethality* of firearms. Also referred to as a clip.

magistrate

a quasi-judicial who presides over minor cases and preliminary hearings.

MAGLOCLEN

Middle Atlantic-Great Lakes Organized Crime Law Enforcement Network. See *Regional Information Sharing System (RISS)*.

mail-order fraud

fraud perpetrated by use of the mail system. Examples of mail fraud include the sale through mail of diplomas.

male prostitute

a man or boy who exchanges sexual acts for money, drugs, or other forms of currency.

malfeasance

the purposeful engaging in misconduct by one in the performance of official duties, generally in the public service. Compare with *misfeasance, nonfeasance*.

malice aforethought

the intent to do wrong that precedes the offense.

malum in se

the Latin term meaning bad in and of itself. In the history of criminal law, some offenses, such as incest and cannibalism, were deemed so reprehensible that they were considered crimes against humanity irrespective of any laws prohibiting them. Examples of offenses mala in se are homicide, rape, and robbery. In most societies, such offenses are subject to severe criminal penalties. Compare with *malum prohibitum*.

malum prohibitum

the Latin term meaning bad because it has been prohibited. An example of a malum prohibitum is drug abuse, which historically and universally has not been deemed inherently wrong. Compare with *malum in se*.

mandatory arrest

a policy in which those accused of domestic violence are automatically arrested. Mandatory arrest policies resulted from the realization that victims of domestic violence remain vulnerable unless the assailant is arrested and jailed.

mandatory minimum

a minimum term of confinement prescribed by law for persons convicted of certain offenses.

mandatory release

the release of a convicted offender from a correctional facility due to *good time* or other statutory provisions.

mandatory sentence

a criminal sentence that must be served. Mandatory sentences became popular in the 1970s and 1980s with the abandonment of the rehabilitative ideal. Because inmates must serve more time than they would under *indeterminate sentencing*, mandatory sentences cause prison populations to burgeon. See *indeterminate sentence, nothing works.*

Manhattan Bail Project

an early large-scale experiment in releasing criminal defendants on their own recognizance instead of having them post cash bail. See *pretrial release.*

manhunt

an organized search by law enforcement or correctional officials for an escaped or otherwise wanted offender. Manhunts can span counties, states, regions, or entire countries. One of the largest manhunts in the United States was that for seven violent offenders who escaped from a Texas prison on December 13, 2000. The manhunt for *Osama bin Laden* in the aftermath of the *Attack on America* was perhaps one of the largest in history.

manic-depression

a mental disorder in which the individual experiences wide mood swings between elevated happiness and depression. The term bipolar disorder has replaced manic-depression as the preferred terminology.

manifest function

the intended or anticipated consequences of law or rules. A manifest function of a law increasing the penalty for armed robbery would be to deter would-be robbers from committing such crimes. See *latent function.*

Mann Act

the Mann White Slave Traffic Act of 1910.

manner of death

the way in which a person dies. The manner of death can differ from the actual *cause of death.* For example, the actual cause of death might be listed as cerebral anoxia and the manner of death as hanging.

Mannheim, Herman

a well-known academic criminologist of the 20th century. Mannheim was affiliated with the London School of Economics. His students included University of Chicago criminologist and law professor Norval Morris.

Mano Nera

same as *Black Hand*.

manslaughter

the nonintentional taking of a human life. Manslaughter is considered less serious than *murder*. See *involuntary manslaughter, voluntary manslaughter*.

Manson, Charles

leader of the so-called Manson Family that was responsible for the 1969 murders of actress Sharon Tate and four guests at her home. The Manson Family later murdered Leno and Rosemary Labianca. He was sentenced to death, but his sentence was later commuted to life imprisonment. Every time Manson has come up for parole, his petition is denied, not only because of the ferocity of his crimes but also because of the efforts of prosecutors and victims' families. See *mass murder*.

marbling

an effect on the human body produced by the hemolysis of blood vessels in reaction with hemoglobin and hydrogen sulfide. Marbling takes its name from the resultant green, purple, and red discoloration of the skin.

marijuana

Cannibis sativa, an organic hallucinogenic drug.

Marion

the site of a U.S. Penitentiary at Marion, Illinois. The facility at Marion has the reputation for housing some of the most dangerous federal convicts.

marital rape

sexual intercourse forced on one spouse by the other.

mark
the person who is the intended victim in a *confidence game* or swindle.

marshal
a law enforcement officer associated with a judicial district.

Marxist criminology
a theoretical perspective in criminology grounded in the writings of Karl Marx. Compare with *conflict criminology*.

mask of sanity
coined by psychologist Hervey Cleckley, this expression has been used to describe the seemingly normal persona of the sociopath. See *antisocial personality disorder, psychopath*.

masochism
the practice of deriving sexual pleasure from having pain inflicted.

mass murder
the killing of four or more people in the same location and during the same event. Usually committed by an individual, but there can be multiple perpetrators. Do not confuse with *serial murder*. See *James Huberty, Marc Lepine, Richard Speck, Charles Whitman*. Compare with *multiple homicide, serial murder, spree murder*.

mass spectrometer
a machine used in forensic laboratories to identify chemical compositions of substances. Mass spectrometers can be used to analyze paint and other samples in the investigation of crimes.

massacre
the killing of many persons in a single incident.

massage parlor
a business ostensibly offering massages but that is a front for *prostitution* services. Massage parlors routinely advertise their services in the classified sections of newspapers.

masturbation
sexual gratification by self-stimulation of one's genitals.

matricide
the killing of a mother by one or more of her children.

maturation
the process resulting in an offender who desists from engaging in crime.

maximum security
a level of security in correctional facilities for inmates who are assaultive or otherwise dangerous or who pose an escape risk. Compare with *close security, medium security, minimum security, trusty security.*

maximum-security prison
a penal institution that houses the most violent offenders. Compare with *supermax prison.*

mayor's court
a local court presided over by a mayor of a town or municipality that hears minor criminal and traffic cases, such as *driving under the influence* and speeding.

McClellan Commission
A U.S. Senate commission convened to investigate organized crime.

McKay, Henry D.
a criminologist of the early to mid-20th century who pioneered the ecological study of crime and social disorder. McKay was affiliated with the Institute for Juvenile Research. See *Chicago Area Project, Chicago School.*

McVeigh, Timothy
a former U.S. Army veteran who perpetrated the *Oklahoma City bombing.* McVeigh was convicted of 160 counts of murder and was executed by *lethal injection.* See *domestic terrorism.*

meat-eater
a corrupt law enforcement officer who aggressively pursues opportunities for *graft.* Compare with *grass-eater.* See *Knapp Commission, police corruption.*

Medellin Cartel

a powerful narcotics trafficking organization based in Medellin, Colombia.

media violence

violent acts portrayed in various print and electronic media.

mediation

the practice of bringing together adversaries for the purpose of discussing and reconciling differences.

Medicaid fraud

fraud perpetrated by physicians, pharmacists, or insurance companies by overbilling or false billing for services.

medical examiner

see *coroner*.

medical marijuana

marijuana used to alleviate the discomfort or other symptoms of a disease, usually cancer. Medical marijuana has been surrounded by controversy because of the drug's general illegal status.

medium security

security level for correctional facilities between minimum and maximum security for inmates who pose few problems but who still require supervision. Compare with *close security, maximum security, minimum security, supermax prison, trusty security*.

Meese Commission

a commission convened in 1986 to investigate pornography.

Megan's law

name for the Sex Offender Registration Act, a 1996 federal law requiring convicted sex offenders to register with local law enforcement upon their release from prison. Megan's law was named for Megan Kanka, a New Jersey girl who was murdered by a convicted sex offender who lived across the street from her.

Menendez brothers

Erik and Lyle Menendez, brothers who murdered their parents in California. The Menendez brothers maintained in their defense that they had been subjected to extreme physical and emotional abuse, primarily by their father. Their trial, which was highlighted by *Court TV*, ended with their convictions. They were sentenced to life in prison. See *parricide*.

mens rea

literally, guilty mind. Criminal intent. Mens rea traditionally is necessary for an act to be considered criminal. See *actus reus*.

mental health court

a *special court* that is designed to ensure the appropriate treatment of defendants afflicted with mental illness. Pioneered in Florida and Washington, mental health courts followed the model set by *drug courts*.

mentally ill offender

a person suspected, charged, or convicted of a crime who suffers from some form of mental illness. Mentally ill offenders pose a special challenge to the criminal justice system, which is ill prepared to diagnose and treat them.

mercy

compassion toward offenders.

mercy killing

the intentional killing of a person for the purpose of alleviating suffering. An example of mercy killing is an elderly man who shoots his terminally ill wife so she will not have to suffer. Compare with *assisted suicide*.

Merton, Robert K.

an American sociologist noted for his oft-cited article, published in 1938, "Social Structure and Anomie," in which he identified five ends to which individuals aspire and the deviant means they sometimes use to achieve them. Merton spent most of his career as professor of sociology at Columbia University. See *anomie theory*.

mesomorph

a body type characterized by muscularity. Mesomorphs were once thought to be aggressive criminals.

meta-analysis

a methodology used to synthesize the results of a large number of evaluative studies. Meta-analysis involves an examination of the statistical conclusions of the studies' results, permitting the analyst to integrate their findings. Applications of meta-analysis in criminal justice include the analysis of treatment literature in corrections.

methadone

a synthetic form of narcotic used as a substitute for heroin in the treatment of addiction. Methadone does not have the same withdrawal effects as heroin.

methadone diversion

a program designed to divert heroin-addicted offenders from regular criminal justice processing. If the offenders complete the program as prescribed, the criminal charges will be dropped. See *diversion.*

methadone maintenance

a program for those addicted to heroin that permits them to take the drug *methadone* under a physician's care to eliminate the severe withdrawal symptoms associated with withdrawal from heroin. Some critics object to such programs because they in essence substitute one drug, albeit a legal one, for another.

methamphetamine

a synthetic derivative of amphetamine. Originally manufactured and controlled by outlaw motorcycle gangs, methamphetamine can be produced in labs small enough to fit into a motor vehicle. The effects of methamphetamine include increased alertness and euphoria and decreased appetite.

midnight basketball

a delinquency prevention program mentioned in the 1994 Crime Bill. Midnight basketball, which was promoted with good intentions, was ridiculed for its simplistic and unsubstantiated approach to delinquency prevention.

military justice

the system of justice in any branch of the armed forces. See *Uniform Code of Military Justice.*

military police

law enforcement authorities of any branch of the armed forces. Military police function in much the same way as their civilian counterparts.

Milken, Michael

an infamous insider trader of the 1980s. Milken, along with Ivan Boesky and others, was convicted and sentenced to serve 10 years imprisonment—a stiff sentence for this type of crime. His sentence was later reduced for his cooperation in other criminal cases. See *insider trading.*

minimum security

a level of security in correctional facilities for inmates who pose the least amount of risk for escape or rule infraction. Compare with *close security, maximum security, medium security, trusty security.*

minority overrepresentation

the disproportionate number of minorities in criminal justice populations, especially those in confinement.

Miranda warning

a warning that law enforcement officers are required to give to arrested suspects. Suspects must be advised that they have the right to remain silent, that any statement they make can be used against them in a court of law, and that they have the right to an attorney and if they cannot afford an attorney one will be appointed before any questioning. The Miranda warning stems from the 1966 U.S. Supreme Court case *Miranda v. Arizona.*

miscarriage of justice

the failure of the criminal justice system to serve the ends of justice. An example of a miscarriage of justice is when an innocent person is found guilty and sentenced. See *convicted innocent, wrongful conviction.*

misconduct in office

any conduct in the performance of official duties that constitutes criminal behavior or a violation of ethical standards. Misconduct can be *malfeasance, misfeasance,* or *nonfeasance.*

misdemeanor

a minor crime generally punishable by a fine or a term of confinement of less than 1 year. Misdemeanors are typically adjudicated in municipal, county, or other lower courts having limited jurisdiction. Compare with *felony*.

misfeasance

the improper performance of public duties. Compare with *malfeasance, nonfeasance*.

missing person

an individual who is either voluntarily or involuntarily absent from his or her usual and expected environs. See *National Center for Missing and Exploited Children*.

mitigating circumstance

a circumstance surrounding the commission of a crime that argues for leniency in sentencing. Compare with *aggravating circumstance, extenuating circumstance*.

M'Naughten rule

an insanity test named for Daniel M'Naughten, who, in 1843, committed a high-profile murder in England. In his defense, M'Naughten claimed he was insane at the time of the crime and consequently should not be held responsible. He was found not guilty by reason of insanity. For years after his trial, those accused of violent crime used the M'Naughten rule as a defense.

MO

modus operandi or method of operation.

mob

a group of persons whose disorderly behavior threatens public order. Also a term used for *organized crime*.

mob justice

the practice by a mob of meting out punishment to one they believe has committed a crime.

mock prison

a prison-like setting in which students or other experimental subjects experience the organizational and interpersonal dynamics of prison life. Perhaps the most famous mock prison was that created by social psychologists at Stanford University in the early 1970s in which students played roles as either guards or inmates. The experiment was controversial in that those playing guards became extremely brutal and those in the inmate role suffered psychological harm.

Model Penal Code

a criminal code developed and recommended by the *American Law Institute* in 1962. The Model Penal Code was considered progressive in its day.

modified index

the *crime index* plus the offense of arson.

modus operandi (MO)

see *MO*.

mole

an undercover operative who collects information to be used in subsequent investigations or prosecutions.

molestation

the making of improper or illegal sexual advances.

molester

one who engages in molestation.

money laundering

the practice of infusing ill-gotten funds into legitimate businesses so they cannot be easily traced by authorities.

Monitoring the Future

a large-scale study of the behaviors, attitudes, and values of U.S. secondary school students. Monitoring the Future includes items on drinking and illegal drug use.

moral panics

the elevation of a social issue to national prominence that previously received little attention.

mores

see *folkways, norms*.

motive

the reason a perpetrator has for committing an offense.

motor vehicle theft

the theft of an automobile, truck, van, or other motor vehicle. Also referred to as *auto theft*. Due to the value of most motor vehicles, such an offense is generally a *felony*. Motor vehicle theft is an *index crime* as defined by the *Federal Bureau of Investigation (FBI)*.

mugger

a street robber who often assaults his victims.

multiple homicide

the killing of many victims by an offender or set of offenders within a short period of time. Multiple homicide can take the form of *mass murder*, *serial murder*, or *spree murder*. Also referred to as multiple murder.

multiple personality disorder

a personality disorder in which the patient evidences two or more distinct personalities, each often with its own background, manner of speech, and so on.

multisystemic therapy (MST)

a comprehensive family-based treatment for antisocial behavior. MST addresses a wide variety of factors at the individual, family, peer, and school levels. Designated a Blueprint program by the *Center for the Study and Prevention of Violence (CSPV)* at the University of Colorado, MST is empirically grounded in the literature on what works. Evaluations of MST have shown it to be highly effective.

mummification

the drying of a corpse due to burial in a dry place or exposure to a dry climate.

municipal court

a lower court serving a municipality and surrounding area that hears misdemeanor cases.

murder

the intentional taking of a human life. Compare with *involuntary manslaughter*, *voluntary manslaughter*.

Murder Castle

term given to the home of notorious serial killer Herman Mudgett, who, in the late 1800s, lured more than 200 female victims to his castle-like residence, where he tortured and murdered them. Mudgett was convicted and executed for his crimes.

Murder, Inc.

a group of organized criminals who, in the 1940s, carried out contract killings for the Mafia.

Murph the Smurf

nickname given to Jack Murphy, who, with an accomplice, burglarized the Museum of Natural History and stole a number of precious gems in 1964.

mutilation

the disfigurement of a person, alive or deceased.

muzzle

the open end of a firearm's barrel from which the bullet exits. Compare with *butt*.

muzzle stamp

the outline on a shooting victim's skin of the front sight and muzzle of a gun barrel. Occurs when the gun barrel is in contact with the skin, and the body cavity immediately beneath the skin allows for the infusion of gases from the discharge of the gun. See *abrasion collar, contact wound, entrance wound, exit wound, starring, strippling, tattooing*.

mystery

a crime novel in which the identity of the perpetrator is not revealed until the end of the book. See *suspense novel*.

naive check forger

forgers who are not professional criminals. Naive check forgers, according to Edwin M. Lemert, are impulsive, nonviolent, and often likable. Compare with *systematic check forger*.

narcissistic personality disorder

a personality disorder characterized by feelings of grandiosity and entitlement. Those with narcissistic personality disorder demonstrate a need for attention and a lack of empathy toward others.

narcotic

a drug (as opium) that in moderate doses dulls the senses, relieves pain, and induces profound sleep but in excessive doses causes stupor, coma, or convulsions.

National Advisory Commission
on Criminal Justice Standards and Goals

a distinguished body of criminal justice experts convened during the early 1970s. The National Advisory Commission issued a series of influential reports on the state of a number of criminal justice functions, including law enforcement, courts, corrections, and information systems.

National Archive of Criminal Justice Data (NACJD)

a repository of crime- and justice-related data sets at the University of Michigan. Part of the *Inter-University Consortium for Political and Social Research (ICPSR)*, the NACJD provides no-cost access to hundreds of criminal justice data sets. The NACJD, which receives financial support from the *Bureau of Justice Statistics (BJS)*, also makes available an online data analysis system for its users.

National Center for the Analysis of Violent Crime (NCAVC)

a division of the Federal Bureau of Investigation's *Critical Incident Response Group (CIRG)*. The NCAVC investigates unusual or repetitive crimes. Its programs include *VICAP (Violent Criminal Apprehension Program)*.

National Center for Health Statistics (NCHS)

a unit within the *Centers for Disease Control and Prevention (CDC)*, U.S. Department of Health and Human Services, responsible for providing statistics that ultimately improve the health of Americans. The NCHS conducts both ongoing and periodic surveys, and it serves as a repository of a vast amount of health-related data, including those related to intentional injury and death.

National Center for Juvenile Justice (NCJJ)

the juvenile justice research arm of the *National Council of Juvenile and Family Court Judges (NCJFCJ)*. Located in Pittsburgh, Pennsylvania, the NCJJ is a nonprofit organization. It engages in applied research, legal research, and systems research.

National Center for Missing and Exploited Children

a center dedicated to locating missing children throughout the United States. The National Center for Missing and Exploited Children was launched in part by the well-known abduction and murder of Adam Walsh on July 28, 1981.

National Center for State Courts (NCSC)

a nonprofit organization founded in 1971 that provides leadership and service to state courts. Issues the NCSC assists courts with include court administration, case flow management, and court technology. Its research agenda has included such topics as community-focused courts, *alternative dispute resolution*, and domestic relations. The National Center for State Courts is headquartered in Williamsburg, Virginia.

National Center for Victims of Crime (NCVC)

a national organization committed to helping crime victims rebuild their lives. In addition to advocating for victim rights and resources, the NCVC offers training and technical assistance.

National Center on Institutions and Alternatives (NCIA)

a nonprofit organization headquartered in Alexandria, Virginia, that promotes humane alternatives for offenders. See *Client-Specific Planning*.

National Commission on the Causes and Prevention of Violence

a commission convened by President Lyndon Johnson in response to the social violence of the late 1960s. Methods for collecting relevant data included citizens surveys and the analysis of archival data.

National Commission on Correctional Health Care (NCCHC)

a nonprofit organization dedicated to improving the quality of health care in detention facilities, jails, and prisons. The services offered by the NCCHC include accreditation, conferences, training, and technical assistance.

National Council of Juvenile and Family Court Judges (NCJFCJ)

an organization dedicated to improving juvenile and family courts in the United States. The NCJFCJ conducts educational and research programs, and it strives to improve the juvenile court system through a variety of means. Research for the NCJFCJ is conducted by the *National Center for Juvenile Justice (NCJJ)*.

National Council on Crime and Delinquency (NCCD)

a nonprofit organization that "promotes effective, humane, fair, and economically sound solutions to family, community, and justice problems." NCCD, established in 1907 as the National Probation and Parole Association, has taken policy positions on such issues as minorities and females in the justice system, gun violence, and the death penalty. In addition to advocacy work, NCCD staff have undertaken a number of influential research and evaluation studies. It maintains offices in San Francisco and Madison, Wisconsin.

National Crime Information Center (NCIC)

a national electronic data center that contains more than 10 million records, including those of stolen autos, stolen license plates, stolen or missing guns, and wanted persons. More than 80,000 law enforcement agencies are part of the NCIC. When a law enforcement officer runs a check on NCIC, it comes back as a "hit" or a "miss." The officer thus knows if the person is wanted and dangerous.

National Crime Prevention Council (NCPC)

a national organization that promotes the reduction of crime through addressing its underlying causes and by reducing opportunities for crime to occur. Much of the NCPC's work relates to a national anticrime publicity campaign, including McGruff the crime dog. The NCPC also holds several conferences each year, some focusing on selected aspects of crime or delinquency prevention. The NCPC is headquartered in Washington, D.C.

National Crime Survey

see *National Crime Victimization Survey*.

National Crime Victimization Survey (NCVS)

a periodic survey of citizens conducted by the U.S. Bureau of the Census for the *Bureau of Justice Statistics*. The NCVS measures the respondents' experiences as victims of rape, robbery, assault, burglary, larceny, and motor vehicle theft. To prevent the identification of respondents, NCVS data cannot be disaggregated. See *self-report study*.

National Criminal Justice Association (NCJA)

a national, nonprofit organization headquartered in Washington, D.C., whose membership is composed primarily of *state planning agencies* and other criminal justice professionals.

National Criminal Justice Reference Service (NCJRS)

a service of the U.S. Department of Justice that makes available a wide range of documents and bibliographical sources regarding crime and criminal justice.

National District Attorneys Association (NDAA)

a nonprofit organization that represents the interests of prosecutors. The NDAA engages in a variety of activities to improve the effectiveness and professional standing of district attorneys.

National Incident Based Reporting System (NIBRS)

the crime reporting program designed to replace the Uniform Crime Reporting program of the *Federal Bureau of Investigation (FBI)*. For each criminal offense reported, law enforcement agencies participating in NIBRS collect detailed data on the offender, time, place, and other aspects of the crime. The advantage of NIBRS over summary-based crime reporting is that users can perform detailed analyses of the correlates of crime. See *Uniform Crime Reports*.

National Instant Criminal Background Check System

a national computerized database used to screen handgun buyers. This system grew out of the Brady Handgun Violence Prevention Act of 1993. The purpose of the instant check system is to enable law enforcement authorities to conduct background checks on those trying to purchase handguns.

National Institute on Drug Abuse (NIDA)

a federal organization that conducts and promotes research on drug abuse and addiction.

National Night Out

an annual crime prevention event sponsored by the National Association of Town Watch. The purpose of National Night Out is to emphasize the right of citizens to enjoy their communities without the fear of criminal victimization. Many local communities participate in National Night Out, sponsoring a variety of community events.

National Opinion Research Center (NORC)

a social science research organization affiliated with the University of Chicago. NORC conducts numerous national and regional surveys, some of which directly or indirectly address crime and justice-related topics.

National Organization for the Reform of Marijuana Laws (NORML)

a national membership organization dedicated to lessening the severity of laws prohibiting the manufacture, cultivation, sale, and especially use of marijuana. See *medical marijuana*.

National Organization for Victim Assistance (NOVA)

headquartered in Washington, D.C., NOVA is a nonprofit organization committed to recognizing and implementing victims' rights and services. Among its achievements are the institution of victim compensation programs and the promotion of victim input in the criminal justice process.

National Organization of Black Law Enforcement Executives (NOBLE)

an organization of black police chiefs who are dedicated to improving law enforcement and the administration of criminal justice.

National Rifle Association (NRA)

a large nonprofit organization with headquarters in Virginia whose mission is to protect the rights of gun owners and sportsmen in the United States. The NRA has traditionally contributed substantial sums of monies to the political campaigns of federal and state politicians whose views align with its views. With respect to crime control, the NRA has emphasized the responsibility of criminals for violent crime. Consequently, it advocates harsher sentences for convicted offenders as the most sensible means for stemming gun-related crime. See *Second Amendment*.

National Sheriffs Association (NSA)

a national organization of county sheriffs. Headquartered in Alexandria, Virginia, the NSA promotes professionalism among sheriffs and assists members in obtaining federal and state funding.

National Youth Gang Survey

a large-scale survey of law enforcement agencies regarding their knowledge of, and experience with, juvenile gangs.

National Youth Survey

a program of self-report studies in which adolescents are interviewed about their involvement in crime and other behaviors related to a delinquent lifestyle.

nature-nurture debate

an old but ongoing debate about whether crime is due to biological causes or environmental causes, including child rearing. Many criminologists now concede that both sets of factors play a role in the *etiology* of crime.

NCAVC

see *National Center for the Analysis of Violent Crime (NCAVC)*.

NCIC

see *National Crime Information Center (NCIC)*.

NCJRS

see *National Criminal Justice Reference Service (NCJRS)*.

necrophagia

the practice of eating the flesh of dead persons. Necrophagia differs from cannibalism in that it involves sexual pleasure.

necrophilia

the aberrant fixation with death and dead persons. Necrophilia can manifest itself in a number of ways, ranging from sexual fetishism associated with coffins and other symbols of death to sexual relations with corpses. Necrophilia generally manifests itself as a criminal offense as abuse of a corpse.

neighborhood watch

an organized effort by citizens to monitor their neighborhood to prevent crime and report suspicious activity. Despite its popularity with law enforcement and citizens, research suggests that neighborhood watch programs have little value in preventing crime.

neoclassical school of criminology

the school of criminological thought that introduced the notion that *free will* could be compromised by mental defects, rendering accused offenders less legally responsible for their behavior. The neoclassical school emerged in the early to mid-1800s with certain high-profile criminal cases in which the offenders claimed insanity.

neonaticide

the killing of a neonate or newborn child. In the late 20th century, a spate of neonaticides were committed by teenage girls who had kept their pregnancies secret from their parents and others.

net widening

the unintended expansion of a service population. Net widening in criminal justice can occur when those to be diverted or otherwise served would not have been part of the original target population. See *diversion*.

neutralization techniques

rationalizations employed by delinquents and other offenders to mitigate the seriousness of their behavior. Neutralization techniques, which were identified by David Matza, include such rationalizations as denial of a victim, denial of an injury, condemnation of the condemners, and denial of responsibility.

New York jack

a slang term for heroin.

new-generation jail

term used to describe a jail that is more progressive in its design and operation. Features of new-generation jails can include the elimination of barriers between staff and inmates and the development of reintegrative programming.

NGRI

see *not guilty by reason of insanity.*

NIBRS

see *National Incident Based Reporting System (NIBRS).*

night prosecutor's program

a program in which prosecutors hear and attempt to settle disputes after normal working hours without the formal filing of criminal charges. Where night prosecutor's programs were available, they tended to emphasize *domestic violence* cases and disputes between neighbors that led to assault. See *dispute resolution.*

nightstick

a long club carried by law enforcement officers. Compare with *billy club PR-24, truncheon.*

no contest

a plea entered by a defendant in criminal court in which there is agreement with the facts as stated in the complaint but no admission of guilt. In most cases, a person who pleads no contest is found guilty by the court. Also called *nolo contendre.*

no-knock law

a law that permits law enforcement authorities to enter a premises without notice to the occupants. Also known as no-knock rule.

nolle prosequi

a disposition of a criminal case in which the prosecutor has decided not to proceed further. In the case of plea bargaining, it is not uncommon for prosecutors to "nolle" one or more charges in return for a plea of guilty to the remaining charges. Nolle prosequi may also be used in cases in which the prosecutor deems the available evidence is insufficient to win a conviction. See *plea bargain.*

nolo contendere

see *no contest*.

nomadic killer

a serial killer who roams from location to location to find and kill victims. Compare with *territorial killer*.

nonfeasance

the failure to carry out official duties. Compare with *malfeasance, misfeasance*.

nonintervention

an expression and philosophy popularized by sociologist Edwin M. Schur in his 1973 book *Radical Non-intervention*. Nonintervention rests on the assumption that the juvenile justice system and other such systems often do more harm than good.

nonlethal weapon

a weapon that can be employed by law enforcement officers against suspects without resulting in death or serious injury. There has been substantial research and development work sponsored by the *National Institute of Justice* and by the U.S. armed forces.

nonpayment of child support

a criminal charge that stems from the failure by a parent or legal guardian to pay child support that was previously ordered by a court. In some states, the nonpayment of child support, also called nonsupport, is a felony and thus punishable by a prison term. Imprisonment does not relieve the negligent parent of his or her responsibilities: If the parent once again fails to pay support, he or she can once again be sentenced to prison.

nonreporting probation

a form of probation in which the probationer does not have to check in with the probation officer. Offenders on nonreporting probation most often are minor offenders or others who are perceived as posing little *recidivism* risk. In such cases, probation officers have little to do other than periodically run record checks for possible violations. See *probation*.

norms

a standard of conduct in society. In most cases, crime violates social norms. Compare with *ethics*.

not guilty

a defense plea expressing that the defendant has not engaged in the conduct in question.

not guilty by reason of insanity (NGRI)

a plea and disposition that states that the defendant cannot be legally responsible for the crime with which he or she was charged due to not knowing right from wrong.

not in my back yard (NIMBY)

a phrase that conveys residents' sentiments against the placement of halfway houses, prisons, and offenders in their neighborhoods. Of particular interest is that many people who in general favor such progressive policies as the rehabilitation of offenders and the community treatment of the mentally ill find themselves saying NIMBY when it comes to their own neighborhood or community.

nothing works

a mantra that conveys the belief that correctional interventions have no rehabilitative value. Nothing works began in the 1970s with the publication of *The Effectiveness of Correctional Treatment*, an evaluation of correctional programs, by Robert Martinson, Douglas Lipton, and Judith Wilks. This evaluation noted that, with few exceptions, the programs had little appreciable value in changing offenders. The conclusion that nothing works in corrections was later challenged by other criminologists who cited evidence to the contrary.

NOVA

see *National Organization for Victim Assistance (NOVA)*.

numbers racket

an illegal lottery game in which people pay to bet on certain sets of numbers in hopes of winning a large sum of money. Often referred to as a poor man's lottery, the numbers racket generally is found in low-socioeconomic neighborhoods in which those with little chance of bettering their financial situation hang their hopes on winning in the numbers racket. These rackets are typically operated by organized crime.

Nuremberg principle

the principle that those faced with carrying out orders should disobey those orders if they are unjust. The Nuremberg principle stemmed from the post-World War II trials of Nazi war criminals, many of whom claimed that their crimes were the result of simply carrying out the orders of superiors. The magnitude and horror of the crimes with which they were charged, however, argued not only for criminal responsibility but also for severe penalties.

obscene phone call

a telephone call made by one whose purpose is to utter obscenities, sometimes for sexual gratification.

obscenity

an obscene act or written or spoken word. Compare with *pornography*.

obstruction of justice

the intentional hindering of an official investigation or subsequent related justice processes.

occult crime

crime perpetrated by a person or persons who practice the occult arts. An example of an occult crime is the practice of human sacrifice by those who hold black masses, worship Satan, or otherwise engage in the black arts.

occupational crime

crime committed by persons during the performance of their legitimate occupation. Examples of occupational crime include *Medicaid fraud* by physicians and pharmacists.

occupational deviance

deviant behavior engaged in during the course of a legitimate occupation. *Scientific misconduct* is an example of occupational deviance.

odontology

see *forensic odontology*.

off-duty weapon

a firearm, most often a *handgun*, carried by a law enforcement officer when not on duty. Compare with *service weapon*.

offender-based transaction statistics (OBTS)

criminal justice data that document the full range of an offender's transactions in the criminal justice system, including arrest, prosecution, trial, sentencing, and corrections. OBTS data permit policymakers to analyze such topics as the effects of proposed or enacted criminal justice legislation. The advent of computerization in criminal justice and the ability of disparate data systems to communicate with each other have facilitated the practical use of OBTS.

Office for Victims of Crime (OVC)

a branch of the U.S. Department of Justice's *Office of Justice Programs (OJP)*. The OVC administers grant programs intended to assist victims and victim advocate organizations in the states and territories.

Office of Community Oriented Policing Services (COPS)

the federal office within the *Office of Justice Programs* of the U.S. Department of Justice charged with promoting community policing. The COPS office oversees several grants programs, including those intended to facilitate the hiring of President Bill Clinton's promised 100,000 community policing officers by local law enforcement agencies. It also funds a number of regional community policing institutes throughout the country.

Office of Justice Programs (OJP)

the branch of the U.S. Department of Justice that administers funding programs. OJP is composed of the *Bureau of Justice Assistance (BJA)*, the *Bureau of Justice Statistics (BJS)*, the *National Institute of Justice (NIJ)*, the *Office for Victims of Crime (OVC)*, and the *Office of Juvenile Justice and Delinquency Prevention (OJJDP)*.

Office of Juvenile Justice and Delinquency Prevention (OJJDP)

a branch of the U.S. Department of Justice's *Office of Justice Programs*. OJJDP is charged with promoting a variety of programs to reduce juvenile delinquency and to improve the administration of juvenile justice. Authorized by the Juvenile Justice and Delinquency Prevention Act of 1974, OJJDP administers discretionary and formula grant programs.

Office of National Drug Control Policy (ONDCP)

the federal office located administratively within the Executive Office of the President charged with developing and overseeing the administration's antidrug policies. The emphasis of the ONDCP is the reduction of drug use, manufacturing, and trafficking. The ONDCP has been the subject of controversy largely because of its emphasis on drug interdiction and *supply reduction*. This is reinforced by the appointment of former military leaders as directors.

Oklahoma City bombing

the 1995 bombing of the Alfred P. Murrah federal office building in Oklahoma City, Oklahoma. The Oklahoma City bombing resulted in 168 deaths and numerous injuries. It is considered the worst instance of *domestic terrorism* on United States' soil. See *Terry L. Nichols, Timothy McVeigh*.

Old Bailey

the trial court in London.

omerta

the oath new members allegedly take upon joining the *Mafia*. Omerta, first introduced to the public by Mafia turncoat Joseph Valachi in his testimony before the *McClellan Commission*, requires that the member swear never to reveal information about the Mafia or else suffer the penalty of death. A ritual including the burning of paper allegedly accompanies the oath.

Omnibus Crime Control and Safe Streets Act of 1968

a comprehensive, influential crime bill passed in the wake of the assassinations and civil disorder of the 1960s. This act provided for millions of dollars in crime control and established the *Law Enforcement Assistance Administration* to oversee it.

OMVI

operating a motor vehicle while intoxicated. Same as *driving under the influence*.

onset

the point at which *criminal careers* begin. See *desistance*.

Operation Borderline

a 1988 sting operation of the U.S. Customs Bureau in which pedophiles were arrested when they ordered *child pornography* from a phony mail-order house.

Operation Brilab

a *Federal Bureau of Investigation (FBI)* sting operation involving organized crime in Louisiana and Texas.

Operation Cease-Fire

a project in Chicago designed to prevent and respond to incidents of gun violence. Participants in Operation Cease-Fire include the Chicago Police Department and numerous community institutions and organizations.

Operation Chaos

a special operations group of the *Central Intelligence Agency (CIA)* whose purpose was to investigate foreign influence in protest activity within the United States. This assignment for the CIA from President Johnson represented a dramatic departure from that agency's foreign focus. Agents amassed files on thousands of Americans. See *political crime*.

Operation Ill Wind

a *Federal Bureau of Investigation (FBI)* operation focused on Department of Defense (DOD) contracts. DOD personnel colluded with defense contractors to defraud the U.S. government of millions of dollars. The investigation led to the conviction of Pentagon officials, contractors, and others.

Operation Mongoose

an attempt by the *Central Intelligence Agency (CIA)* to enlist organized criminals to assassinate Cuban Premier Fidel Castro.

Operation Underworld

a U.S. Navy program during World War II in which the Navy enlisted the assistance of the mob to protect U.S. docks.

Operation UNIRAC

a *Federal Bureau of Investigation (FBI)* sting operation investigating organized crime connected to longshoremen and shipping companies. The operation demonstrated the influence of organized crime in these industries.

Opium Wars

conflicts between the Chinese and Europeans from 1839 to 1842 over the ability to trade in opium.

opportunity theory

the notion in criminology that some offenders engage in crime simply because of the opportunity to do so.

Oraflex

a drug marketed by the Eli Lilly Company in 1985 despite knowledge that it had deadly side effects. See *corporate crime*.

organic brain dysfunction

brain damage that can influence an individual's susceptibility to engage in criminal behavior.

organizational crime

crime perpetrated with the knowledge, consent, and sometimes active participation of high-level officials within an organization.

organizational deviance

deviance engaged in by employees of an organization on behalf of the organization.

organized crime

crime perpetrated by a formally organized syndicate or loosely organized collectivity. See *Mafia*.

organized offender

a type of serial killer whose crimes give evidence of a more methodical, planned offense. Organized serial killers tend to be more intelligent, better educated, and more apt to have normal relationships with members of the opposite sex than their disorganized counterparts. They often kill their victims at a location different from where they are found. See *disorganized offender, serial killer*.

other-report study

a form of criminological study in which the researcher asks subjects about the criminal or deviant behavior of others. Compare with *self-report study*.

outlaw motorcycle gang

an organized group of motorcyclists whose activities center around criminal enterprises. Examples of outlaw motorcycle gangs are the *Hell's Angels* and the Outlaws.

outsiders

a term coined by sociologist Howard Becker in his book by the same name to describe those who, on being labeled by society and therefore denied the legitimate means to carry on normal lives, turn to illegitimate activities.

overcriminalization

the excessive reach of criminal law resulting in defining relatively minor or harmless acts as criminal.

overcrowding

the state wherein the number of jail or prison inmates exceeds the reasonable capacity of the correctional system to house or otherwise care for them. Also referred to as crowding. See *design capactiy*.

overdose

an amount of a drug that exceeds the recommended or safe dose. An overdose of a drug can kill or cause serious illness or injury.

overkill

the practice of continuing to shoot, stab, or bludgeon after death has obviously occurred.

Palestine Liberation Organization (PLO)

an organization of Palestinian nationalists dedicated to the liberation of Palestine. The PLO maintains the right to use violence as a means to achieve its ends. See *terrorism*.

panopticon

a prison whose design is credited to *Jeremy Bentham* in which multitiered cells are built around a hub so that correctional personnel can view all inmates at the same time.

pansexual

one who engages in diverse sexual relations.

paper hanging

the purposeful execution of checks or similar instruments knowing there are insufficient funds to cover them. See *forgery, passing bad checks*.

paper trail

the chain of documents that links an offender to the offense, especially in financial crimes such as embezzlement.

paraffin test

a test used by criminalists to determine if an individual has recently discharged a firearm. When someone fires a firearm, especially a handgun, minute particles of gunpowder adhere to the skin on the hand and arm. The paraffin test detects these particles.

paramilitary

a group organized along the lines of a military unit.

paraphilia

any of several sexual deviations, including but not limited to exhibitionism, pedophilia, sexual sadism, and necrophilia. Whereas some paraphilias are *victimless crimes*, others represent harmful behavior.

pardon

an official decree by a president, governor, or other high-ranking official that sets aside any remaining criminal penalties. One of the most controversial uses of the pardon was by President Bill Clinton, in which he pardoned on his final day in office.

parens patriae

the philosophy that juvenile justice and other youth-serving officials serve in the place of parents. Parens patriae emerged in response to the harsh treatment that children have received in institutions and work settings. In the latter part of the 20th century, this philosophy was called into question, with a movement toward harsher treatment of juvenile offenders.

parent abuse

the intentional battery or other physical abuse of a parent by a child. Parent abuse often takes the form of *elder abuse.*

parenticide

the killing of one's parents. See *parricide*.

Parents of Murdered Children (POMC)

the full name is Parents of Murdered Children and Other Survivors of Homicide Victims, often abbreviated POMC. This national organization, headquartered in Cincinnati, Ohio, is dedicated to providing emotional support to those who have lost loved ones to homicide. There are many local chapters throughout the country, each of which attempts to help members cope with the tragic aftermath of murder. POMC depends primarily on charitable contributions for its operating funds. It also provides training to professionals whose work is touched by homicide.

parole

the conditional release under supervision of a convict prior to the expiration of the sentence. Parole is generally granted by a *parole board*. Upon release on parole, the offender periodically reports to a *parole officer*, who ensures that the *parolee* abides by certain conditions. Compare with *aftercare*.

parole authorities
the state or federal organizational unit that oversees the administration of *parole*.

parole board
a body composed of professionals and laypersons who make recommendations and decisions regarding the suitability of release on parole.

parole officer
an officer charged with supervising and monitoring the whereabouts of persons on parole. Parole officers may also prepare parole violation reports, and they may verify placements for those eligible for release from prison.

parole violation
a violation of the conditions of release on parole, including the commission of a new offense.

parole violator
a person on parole who has committed either a new offense or a *technical violation*.

parolee
a person released from *prison* on *parole*.

parricide
the killing of a relative, usually of a parent. See *parenticide*.

participant observation
a research methodology in which the researcher mingles with the subjects of the study. Laud Humphreys' study of *tearoom trade* was a participant observation study.

Part I offense
any of eight offenses included in *Uniform Crime Reports'* crime index. Part I offenses include *murder* and nonnegligent homicide, *rape, robbery, aggravated assault, burglary, larceny, motor vehicle theft,* and *arson*. Also referred to as Part I crime. Compare with *Part II offense*.

Part II offense

any offense reported to law enforcement not included as Part I offenses. Part II offenses include, but are not limited to, *forgery, passing bad checks, petit theft,* and *driving under the influence.*

passing bad checks

the issuance of checks when the issuer knows there are insufficient funds to cover them. Compare with *forgery.*

pathologist

a physician who specializes in determining the causes of death due to disease or injury. Pathologists examine organ structure and human tissue to uncover the circumstances surrounding death. For example, a pathologist in one case found bruising of the neck muscles consistent with death by strangulation, despite the assertions by the victim's boyfriend that the woman had died from drowning.

pathology

the science of diseases.

pathways to crime

the various ways in which a youth moves from unruly to overtly illegal behavior. Research has shown that there are distinct pathways to crime.

patricide

the killing of one's father. One of the most famous patricides was the killing by Eric and Lyle Menendez of their father (and mother) in California, who reportedly was demanding and abusive.

patrol officer

same as *patrolman.*

patrol wagon

a police van capable of holding several arrestees. Also referred to as paddy wagon.

patrolman

a uniformed police officer whose responsibility is to patrol a specified geographical area. See *patrol officer.*

payola

a bribe offered in return for a favor by a politician or other official.

peacemaking criminology

a school of criminological thought emphasizing empathy, understanding, and cooperation in addressing the problems of crime and delinquency.

peculate

to embezzle. See *embezzlement*.

pedophile

one who engages in *pedophilia*.

pedophilia

a sexual attraction to children.

Peeping Tom

a voyeur.

penal colony

a separate tract of land, often on an island or other remote location, to which condemned prisoners are assigned. One of the most infamous penal colonies was *Devil's Island*, the name for a penal colony in French Guiana.

penal institution

same as *prison*.

penal philosophy

preference regarding the punishment and treatment of offenders.

penal reform

organized efforts to lessen the severity or inhumanity of criminal sanctions, especially imprisonment.

penal sanction

any legal consequence upon criminal conviction, especially those involving confinement or control.

penal servitude

serving as an indentured servant as part of a penal sanction.

penitentiary

a correctional facility designed for the long-term confinement of convicted adult felons.

Pennsylvania school of criminology

a tradition of criminological research originating at the University of Pennsylvania. The most influential figures of the Pennsylvania school of criminology were *Thorsten Sellin* and *Marvin E. Wolfgang*.

Pennsylvania system

a system of corrections characterized by solitude, hard work, and reflection for the inmates.

penology

the study of punishment and corrections.

Pentagon procurement scandal

a scandal in which the U.S. armed forces paid exorbitant prices for goods in return for *kickbacks* from the contractors.

pepper spray

an aerosol containing capsicum that is used for personal protection. Pepper spray causes severe irritation of the eyes and breathing difficulties, rendering the person sprayed temporarily helpless. See *mace*.

peremptory challenge

the right during *voir dire* to challenge the seating of jurors without citing a specific cause.

perjury

lying under oath in a court of law. Perjury is a *felony* punishable by imprisonment.

perpetrator

one who has committed or is suspected of committing a crime. Also perp.

persistence

the lasting quality of *criminal careers*. See *desistance, onset*.

personal crime
a crime in which the offender must confront the victim.

personal larceny
same as larceny. See *theft*.

perversion
an unnatural proclivity toward a certain behavior, especially of a sexual nature. *Pedophilia* is an example of a perversion.

petechial hemorrhage
pinpoint hemorrhages of the eyes indicative of death by strangulation.

petit jury
a jury, generally of 12 persons, whose role is to listen to the facts of a case in a trial and make a decision. See *grand jury*.

petit theft
theft of goods or services of an amount less than that required to qualify for a felony. Petit theft is a *misdemeanor*.

petty theft
same as *petit theft*.

phony accident claim
see *insurance fraud*.

phrenology
the outdated study of the human skull to determine character, mental ability, and the propensity for criminal activity. Phrenologists measured human skulls, paying particular attention to bumps, indentations, and other peculiarities. See *Bertillon classification system*.

physical evidence
tangible materials that offer the potential of solving a crime and bringing those responsible to justice.

physical stigmata
any physical characteristic or abnormality considered indicative of criminality. Examples of physical stigmata include protruding brow or eyes set closely together.

physical trace evidence

same as *physical evidence*.

pickpocket

a theft offender whose method of operation consists of reaching into the pockets of victims to misappropriate wallets and other goods. Pickpockets employ a variety of tactics, including bumping into their victims.

pigeon

the target of a swindle or con. Pigeons have the reputation of being gullible, if not stupid, for being so easily duped by their victimizers. See *confidence game*.

pigeon drop

a *confidence game* in which the mark is convinced to deposit a sum of money to share in a larger pot. As with most confidence games, the pigeon drop depends in large part on the greed of the mark.

pillory

a wooden structure with holes for the head and hands used to punish and humiliate minor violators in Colonial times. Compare with *stocks*.

pimp

a man who controls a *prostitute* and lives off her earnings. Pimps are more common with street prostitutes, who, often as young *runaways*, are taken in by the pimp's promises of protection and material goods.

PINS

Person in need of supervision—an expression used to describe youths who are not delinquent and as such do not require juvenile justice processing but who need supervision and services. See *status offense*.

piquerism

a sexual proclivity toward the cutting, stabbing, puncturing, or tearing of human flesh. Piquerism often involves intense sexual gratification from engaging in such practices, including the reactions of the victim suffering such trauma, that sometimes result in death.

PIRA

acronym for the Provisional Irish Republican Army.

piracy

the act of robbing ships at sea or aircraft during flight. See *air piracy, highjacking*.

pirate

one who engages in piracy.

pistol

a small firearm designed to be held and fired in one hand.

plagiarism

passing off another's written work as one's own. Plagiarism also includes the theft of ideas, in addition to their manifestation. Although plagiarism generally is dealt with administratively, it can be the subject of criminal or civil proceedings. See *scientific misconduct*.

plainclothesman

same as *detective*.

plea

a defendant's formal, recorded denial or admission to a charge in criminal court.

plea bargain

same as *plea negotiation*.

plea negotiation

the process and result of an agreement between a prosecuting attorney and defense counsel to reduce the seriousness or number of charges in a criminal case in return for a guilty plea. Also referred to as plea bargain.

poacher

one who engages in poaching.

poaching

the illegal capture or killing of game or protected animals. Poaching can be extremely lucrative, so much so that poachers have been known to use violence toward enforcement authorities who try to stop them.

poisoning

the intentional act of administering toxic substances to another person for the purpose of harming or killing them.

police

officials whose responsibility is to enforce criminal laws and ensure public safety.

Police Athletic League (PAL)

a program in which police officers work with youth in athletic and other recreational activities to reduce delinquency and develop responsible citizenship and respect for the law. Originally oriented toward athletics, many PALs now include such activities as computer skills and homework.

police brutality

the application of excessive force by police on citizens. Police brutality received national attention with the beating of motorist *Rodney King* by Los Angeles police officers.

police corruption

the acceptance of bribes and related practices by police. See *internal affairs, Knapp Commission.*

police diversion

a *diversion* program administered by a law enforcement agency.

Police Executive Research Forum (PERF)

a national organization of police executives dedicated to improving policing and advancing professionalism through research and involvement in public policy debate. PERF depends primarily on government and private grants and contracts for funding. Projects that PERF has undertaken include analyses of *racial profiling, community policing, domestic violence,* and *firearms trafficking* as well as the evaluation of *gun buy-back programs.* PERF maintains collaborative relationships with the *International Association of Chiefs of Police (IACP),* the *Police Foundation,* and other law enforcement organizations.

Police Foundation

a nonprofit organization dedicated to improving policing through research, technical assistance, and technology. The Police Foundation, founded in 1970, has been a leading force in promoting community policing through its research. It maintains its headquarters in Washington, D.C.

police misconduct

legal or ethical infractions committed by police. Examples of police misconduct include the use of excessive force, the acceptance of bribes, and the commission of perjury. See *police corruption.*

police officer

a member of a *police* force.

police protection

security services offered by law enforcement authorities to witnesses or others whose lives may be in danger. See *Federal Witness Protection Program.*

political crime

crimes that undermine or otherwise threaten the authority or stability of a government. Political crime can be perpetrated by either domestic or foreign offenders. See *Watergate.*

political espionage

espionage conducted for political purposes. The *Watergate* break-in is an example of political espionage.

political policing

the secret policing practices used by law enforcement authorities to control or suppress unpopular or alleged subversive viewpoints or activities.

political prisoner

a person being held prisoner primarily for ideological reasons.

political terrorism

see *domestic terrorism.*

polygamy

the illegal practice of marrying more than one person.

polygraph test
same as *lie detector*.

population
a large group of individuals under study about which a researcher wants to draw certain conclusions. Compare with *sample*.

pornography
sexually explicit or other objectionable written material, film, video, or photographs.

positive identification
the verification of the identity of a suspect, witness, missing person, or other individual of interest to authorities.

positive school of criminology
the school of criminological thought that emphasized the identification of individual factors responsible for crime. The positive school had its origins in Italy, where *Cesare Lombroso* and *Enrico Ferri* strove to isolate physical features most closely associated with criminals.

positivism
an intellectual movement beginning in the late 19th century emphasizing the collection and analysis of empirical data as well as the diagnosis and treatment of pathological conditions.

possession
the charge for having a quantity of prohibited drugs.

postconviction remedies
any of various legal means of reexamining and changing a criminal sentence.

postmodern school of criminology
a school of criminological thought that challenges dominant perspectives and emphases.

postmortem
occurring after death. Also the medical examination of a deceased after death. Also referred to as *autopsy*.

postmortem lividity

a discoloration of human tissue caused by the effects of gravity pulling the blood to the lowest parts of the deceased. Postmortem lividity can be used to establish if the deceased has been moved after death. Also *lividity*.

postpartum defense

a defense that has been used by some women when charged with certain violent crimes, including the murder of their newborn children. The postpartum defense asserts that during this period physical, mental, and emotional changes in the mother prompt erratic behavior that, in normal circumstances, she would not engage in.

postrelease control

a form of supervision for those released from prison. Postrelease control differs from *parole*.

posttest

a subsequent analysis of a phenomenon under study.

posttraumatic stress disorder (PTSD)

a serious mental disorder experienced by those who have suffered severe trauma. Symptoms of PTSD include nightmares, flashbacks, and depression. PTSD has been used as a defense by those accused of violent crime, including murder. Victims of violent crime may suffer from PTSD.

power elite

sociologist C. Wright Mills's term for those who have the political and financial control in society.

power-control theory

a theory of delinquency that asserts that mothers occupying traditional roles exercise more control over their daughters than their sons. This control keeps the daughters from engaging in delinquency. Mothers occupying untraditional roles do not have such control over their daughters, who are at greater risk of engaging in delinquency.

praxis

in Marxist criminology, action to effect meaningful change.

precedent

a previous court decision on which a current case may rely for authority.

precipitation hypothesis

the assertion that violence as reported and portrayed by the various media actually spawn violent behavior in society. Compare with *catharsis hypothesis*.

predator

an offender who actively stalks or otherwise pursues his or her victims.

predatory violence

violence that is the result of purposeful, premeditated action. Compare with *affective violence*. See *stalking*.

prediction scales

statistics-based instruments used to predict the likelihood of future offending by individuals.

prediction table

a device used by correctional authorities to predict the future risk of recidivism by past behavior and personal characteristics.

preliminary hearing

a criminal court hearing during which evidence is presented by the prosecution in an effort to establish whether to proceed with felony charges.

premeditation

planning a crime in advance in such a way as to show prior intent.

presentence investigation

an investigation undertaken, generally by a probation officer, to supply background information on a defendant prior to sentencing. Presentence investigations typically include family history, employment record, military history, and information about the instant offense, including statements from victims. See *presentence investigation report (PSI)*.

presentence investigation report (PSI)

a report prepared in advance of sentencing, often by a probation officer or investigator, to assist the judge or magistrate in determining an appropriate sentence. See *presentence investigation*.

presentence report

same as *presentence investigation report*.

President's Commission on Law Enforcement and the Administration of Justice

a national commission appointed by President Lyndon B. Johnson in the mid-1960s. Convened in the wake of President Kennedy's assassination and certain civil disturbances during this period, the President's Commission summarized or commissioned research on a number of topics, such as police behavior. This commission's recommendations were instrumental in the passage of the *Omnibus Crime Control and Safe Streets Act of 1968*.

presumptive sentencing

a sentencing scheme in which the legislature has specified a standard sentence, but the sentencing judge may use *discretion* to deviate from that sentence if there are aggravating or mitigating circumstances.

pretest

in research, the administration of a survey, test, or other measure before an intervention or treatment is administered. Compare with *posttest*.

pretrial detention

confinement in a jail or other holding facility prior to trial.

pretrial discovery

see *discovery*.

pretrial diversion

the *diversion* of those charged with crime before a trial has taken place.

pretrial hearing

a court hearing at which motions are made or *plea negotiations* are considered. Compare with *arraignment, preliminary hearing, trial*.

pretrial publicity
media coverage that potentially threatens the ability of the defendant to receive a fair trial. See *change of venue*.

pretrial release
the practice of conditionally releasing criminal defendants prior to trial without the formal posting of bail. Those who participate in pretrial release generally must have stable residence and employment as well as other ties to the community that suggest they are likely to appear for trial. Compare with *bail*.

Pretrial Services Resource Center
a Washington, D.C.-based, nonprofit organization that provides information on pretrial issues for criminal justice professionals and others. The Pretrial Services Resource Center works to improve the criminal justice system at the pretrial stage to improve safety, services, and appearance rates and reduce recidivism. Its services include serving as an information clearinghouse, providing technical assistance, and promoting best practices.

preventive detention
the holding of a defendant prior to trial because of a perceived risk of reoffending or flight.

preventive patrol
patrol by law enforcement officers for the express purpose of preventing crime by maintaining visibility rather than simply reacting to citizens' calls for service.

price-fixing
an illegal collusion between two or more corporations to set the prices of merchandise or services to minimize competition and maximize mutual profits.

primary crime scene
the crime scene in a homicide case in which the body is found. Compare with *secondary crime scene*.

primary deviance
in labeling theory, the conduct that originally caused the labeled person to come into contact with authorities. See *labeling theory*, *secondary deviance*.

primary prevention

the prevention of delinquency before the onset of behaviors characteristic of delinquency. Compare with *secondary prevention, tertiary prevention*.

prior record

at any given point in time, the history of arrests and convictions of an individual. Prior record is often used by criminal courts as one criterion to determine appropriate sentences for defendants.

prison

an institution under state or federal jurisdiction for the confinement of convicted felons.

prison camp

a correctional facility located in a rural area whose physical facilities consist of barracks, tents, or similar temporary construction. Prison camps may also have special purposes for inmates, such as outdoor work.

prison crowding

the condition in which the number of inmates in a prison exceeds an established standard. The issue of prison crowding came to prominence in the United States in the 1970s and 1980s after changes in sentencing laws limited or eliminated parole and kept offenders in prison longer with *determinate sentencing*. Also referred to as *prison overcrowding*.

Prison Fellowship Ministries

an organization founded by *Watergate* figure Chuck Colson. Prison Fellowship Ministries, headquartered in Merrifield, Virginia, promotes the spiritual healing of offenders, their families, the victims, and entire communities.

prison industrial complex

the extensive financial and political enterprise represented by the correctional system.

prison industry

manufacturing facilities operated by correctional systems that employ inmate labor.

prison overcrowding

see *overcrowding, prison crowding.*

prison reform

organized efforts to improve the conditions of prison and assert the rights of prisoners.

prison riot

an often violent uprising by inmates within a prison. There are many causes of prison riots, including abuse by correctional staff and substandard living conditions. Notable prison riots include those that occurred at Attica State Prison in 1971 and New Mexico State Prison in 1980.

prison ship

in past centuries, a ship on which convicted prisoners were held.

prison violence

violence that takes place within a prison. Although prison violence generally conveys violence by inmates against guards or against one another, it can also encompass violence perpetrated by *corrections officers* against inmates.

prisoner

a person deprived of liberty, especially one confined in a prison or jail.

prisoner rights

a broad range of rights of those confined in prison regarding safety and security, health care, freedom from abuse by staff, the ability to exercise, and access to legal materials.

prisonization

generally, the involuntary adoption of the routine and values of a prison as a preferred existence by those so confined.

private security

protection services provided by a nongovernmental firm. Private security provides a substantial amount of public safety services in the United States. See *privatization.*

privately commissioned presentence investigation reports

presentence investigation reports prepared by someone outside of government. Privately commissioned presentence investigation reports became popular with the realization that government presentence investigation reports contained unverified information and did not advance the interests of the person being sentenced. See *Client-Specific Planning.*

privatization

the assumption of criminal justice roles or functions traditionally the province of government by private companies. Privatization in criminal justice has occurred extensively in policing and corrections.

pro bono

performed for the good of the public. Often used to describe donated legal services, including those provided for indigent defendants.

proactive policing

a style of policing in which officers do not simply react to calls for service but instead seek opportunities to prevent crime, apprehend criminals, and otherwise serve the community. Compare with *reactive policing.*

probable cause

a standard that requires that a reasonable individual would believe that a person had committed a crime.

probation

the suspension of a sentence of a convicted offender and granting of freedom for a period of time under specified conditions. Probation, which generally is granted in lieu of confinement, is often confused with *parole.*

probation contract

an agreement between a probation officer and a probationer specifying that certain conditions be met for certain considerations.

probation officer

an officer of a court or state agency who conducts *presentence investigation reports* and supervises convicted offenders on *probation.* See *presentence investigation.*

probation violation

an infringement of the rules of probation that can result in the revocation of probation and the reinstatement of the suspended sentence. A probation violation can stem from a new criminal charge or from a *technical violation*.

probation violator

one who has committed or is suspected of having committed a *probation violation*. Probation violators may be brought before the court, which can result in revocation of probation and reinstatement of the suspended sentence.

probationer

a convicted offender on probation. Probationers must abide by certain conditions imposed by the sentencing judge.

problem-oriented policing

a philosophy and practice of policing that emphasizes the systematic analysis of neighborhood problems and thoughtful solutions. Problem-oriented policing is most closely associated with Herman Goldstein, author of *Problem-Oriented Policing* (1990).

professional criminal

one who makes his or her living by committing theft. Professional crime is distinguished from occasional crime in the identity of the offender with the criminal enterprise as his or her occupation.

professional thief

a person who makes his or her living stealing. *Edwin H. Sutherland* conducted an in-depth study of this type of offender in his book, *The Professional Thief* (1937).

profiler

one who performs *profiling*.

profiling

the practice of using past empirical data on offenders and their behavior to identify, track, or apprehend other such offenders. The *Behavioral Science Unit* of the *Federal Bureau of Investigation (FBI)* has refined the practice of profiling in serial homicide, arson, sexual assault, and other such crimes. Profiling, however, has come under fire in recent years. Many offenders who are stopped by law enforcement authorities fit a racially or culturally based profile. Recent cases have involved black citizens or Arab tourists who are mistaken for suspects. The state of Maryland recently enacted legislation prohibiting the use of profiles. See *racial profiling*.

Program on Human Development in Chicago Neighborhoods (PHDCN)

a well-funded, large-scale research project designed to assess the many causes of delinquency, crime, substance abuse, and violence. The PHDCN research includes individual, family, and community measures. The Harvard School of Public Health directs the PHDCN. See *collective efficacy, cumulative disadvantage, social capital*.

Prohibition

the period in the United States from 1919 to 1932 when the manufacture, distribution, sale, and consumption of alcoholic beverages were prohibited by the *Volstead Act*. Prohibition created a *black market* for alcohol; as a result, organized crime figures such as *Al Capone* met the demand with illegally produced alcoholic beverages, spawning a wave of related corruption and violence. See *bootlegger*.

PROMIS

see *Prosecutor's Management Information System (PROMIS)*.

promiscuity

engaging in sexual relations indiscriminately with multiple partners. Promiscuity, as well as ignorance and unprotected sex, was in part responsible for the AIDS epidemic that began in the 1980s.

property crime

a criminal offense that involves the theft or destruction of property. Examples of property crime are *motor vehicle theft* and *arson*.

prosecuting attorney

an attorney who pursues criminal cases on behalf of a governmental subdivision, such as a county or city.

prosecution

the initiation and pursuit of criminal charges.

prosecution witness

a witness whose testimony is sought to support the legal case made by the *prosecutor*.

prosecutor

see *prosecuting attorney*.

prosecutorial discretion

the ability of a prosecutor to make decisions about whom to formally charge, what to charge them with, and which charges will be reduced or dropped. As with other discretion in the criminal justice system, prosecutorial discretion can sometimes result in unintended miscarriages of justice.

Prosecutor's Management Information System (PROMIS)

an early management information system developed for the scheduling of federal criminal cases, the management of witnesses, and the evaluation of prosecution services. PROMIS served as a model for many of the current information systems in criminal justice.

prostitute

a person who earns money, drugs, or other currency in exchange for sexual acts.

prostitution

the practice of engaging in sexual relations for money. See *prostitute*.

protection extortion

see *protection racket*.

protection order

a judicial order intended to keep a threatened or vulnerable individual safe from another.

protection racket

an illegal enterprise in which business owners are encouraged to pay for protection against robbery, assault, arson, or other crimes that could negatively affect their businesses. Those selling such protection send the message that such crimes will occur if the payments are not made. Also *protection extortion*.

protective factor
any positive attribute of a person at risk of offending that helps insulate against *criminogenic factors*.

prowl car
a police cruiser that patrols neighborhoods searching for crime.

PR-24
a nightstick fashioned of synthetic material with a handle attached perpendicular to the main shaft. PR-24s, which offer advantages over traditional nightsticks, require specialized training. Compare with *baton, billy club, nightstick*.

prurient interest
an unnatural obsession with sexual matters.

psychological autopsy
The art and science of analyzing the personality and mental condition of a person after his or her death.

psychopath
a now dated term for a person who evidences antisocial personality disorder. See *antisocial personality disorder, sociopath*.

psychopathic killer
one who kills, sometimes repeatedly, with no apparent remorse.

psychopathic personality
see *antisocial personality disorder, psychopath*.

psychopathology
the science and study of mental disorders.

psychopathy
a personality characterized by few inhibitions or feelings of guilt. Compare with *antisocial personality disorder*.

Psychopathy Check List-Revised (PCL-R)
a psychometric instrument developed by psychologist Robert D. Hare to assess the extent to which an individual possesses antisocial tendencies and traits. The PCL-R should be completed by a trained professional.

psychosurgery

brain surgery employed to treat mental disorders. Psychosurgery includes but is not limited to lobotomies and lobectomies.

PTL scandal

a scam perpetrated by evangelists Jim and Tammy Bakker of the PTL organization, named for their stock phrase "praise the Lord." Using television and live appearances, the Bakkers solicited millions of dollars in donations from their loyal followers. Instead of using the donations for religious purposes, the Bakkers diverted them for their own use. Both were charged and convicted on federal tax charges.

public defender

an attorney employed by state or local governments whose role is to represent indigent defendants accused of crimes. Compare with *appointed counsel*.

public enemy

a dangerous, notorious criminal. At one time, the *Federal Bureau of Investigation (FBI)* nominated a notorious wanted felon as its Public Enemy #1.

public housing

government-subsidized residences for low-income families. Public housing projects are often plagued by violence, drug trafficking, gang activity, and other forms of crime and social disorder.

public intoxication

drunkenness in public. Similar to drunk and disorderly. See *public order crime*.

public opinion

the feelings and preferences of citizens, often gauged by telephone surveys. Survey organizations such as the *National Opinion Research Center (NORC)* routinely conduct surveys of public opinion, including opinions about crime and justice-related issues.

public order crime

minor offenses that cause more nuisance than harm to society. Examples of public order crimes are disorderly conduct, public drunkenness, and vagrancy. See *public intoxication*.

pugilistic attitude
a commonly found position of those burned to death that resembles a boxer with arms extended as if engaged in a boxing match. Pugilistic attitude results from the contraction of muscles due to exposure to the fire's heat.

Pulse Check
a periodic data collection effort of the *Office of National Drug Control Policy (ONDCP)* to keep abreast of current trends in drug use in the United States. Pulse Check, conducted by *Abt Associates*, employs telephone interviews of experts who have knowledge of substance abuse.

punishment
a penalty meted out to a convicted offender.

purge fluid
a product of decomposition that may drain from the mouth or nose of a deceased person.

pursuit
efforts made by police, most often in motor vehicles, to apprehend a traffic offender or suspect in a crime. See *high-speed pursuit*.

putrefaction
the decomposition of a corpse.

pyramid scheme
an illegal money-making scheme in which converts purportedly make money by enlisting others who pay them residual fees.

pyromania
the pathological desire or compulsion to set things on fire. See *fire setting, serial arson*.

pyromaniac
one who suffers from or engages in *pyromania*. Also *fire setter*.

Quantico, Virginia

the site of the Federal Bureau of Investigation (FBI) National Academy on the U.S. Marine Corps base at Quantico. All FBI agents initially train at Quantico. It is also the site of several important sections, including a crime laboratory, the *Child Abduction and Serial Killer Unit (CASKU)*, and its critical response team. See *Federal Bureau of Investigation (FBI)*.

quasi-experimental design

an evaluative design in which the experimental and control groups are not randomly selected or in which they are not identical. Quasi-experimental designs are employed in criminological research because experiments using true random assignment are seldom practicable in criminal justice settings. See *Academy of Experimental Criminology*.

Quetelet, Adolphe

a 19th-century Belgian statistician who was one of the first to analyze crime statistics to discern regularities and patterns.

qui tam suit

an action taken by an individual against a contractor of the U.S. government alleging criminal or unethical conduct. Qui tam suits were made possible by the Federal Civil False Claims Act (1863, amended in 1943 and 1986).

racial profiling

the practice by law enforcement officials of stopping minorities when no offense has been committed or no reasonable suspicion exists. Racial profiling has garnered much attention, especially with the admission by prominent black celebrities that they have been the object of such police behavior.

racist

one who believes in the superiority of a particular race and hates members of specific racial, ethnic, or religious groups. See *hate crime, Ku Klux Klan.*

Racketeer Influenced and Corrupt Organizations (RICO)

a federal statute that allows law enforcement authorities to charge or civilly sue criminal enterprises. RICO contains several provisions that facilitate the investigation and prosecution of organized criminal activity.

racketeering

obtaining money through a variety of illegal enterprises, including but not limited to *hijacking, prostitution, narcotics,* and *loan sharking.*

radical criminology

a school of theoretical criminology that emphasizes that laws are made by powerful groups in society and that such laws are used to control certain groups. See *Marxist criminology.*

radical nonintervention

see *nonintervention.*

Ramsey, JonBenet

a young girl and regular beauty pageant contestant from a prominent Denver, Colorado, family who was murdered in her home in December 1997. Despite a lengthy investigation, including the consultation of nationally known forensic and behavioral experts, the crime remains unsolved.

RAND Corporation

acronym for research and development; a private research and development corporation based in Santa Monica, California, that conducts criminal justice research and evaluation. RAND has maintained an active criminal justice research program since the early 1970s. RAND's research contributions include studies of *probation, crime prevention* programs, and *three strikes and you're out.*

randomization

in social experiments, the allocation of subjects between experimental and control groups such that each has an equal probability of assignment to either group. See *Academy of Experimental Criminology, experimental design.*

ransom

money demanded by kidnappers and sometimes paid by the victim's family for the safe release of someone from captivity.

ransom note

a note left by kidnappers spelling out their demands.

rape

sexual intercourse by force, threat of force, or deception. See *statutory rape.*

rape kit

a kit containing products to facilitate the collection and preservation of semen, hairs, and other physical evidence taken from a rape victim. Rape kits are typically used by doctors and nurses working in emergency rooms where rape victims are treated.

rated capacity

the capacity of a prison, jail, or other correctional facility based on standardized criteria. Compare with *design capacity.*

reactive policing

a style of policing in which officers respond to calls for service. Compare with *proactive policing.*

reasonable doubt

doubt that a reasonable person would have regarding the alleged guilt of an accused person being tried for a crime. A jury must find the defendant guilty beyond reasonable doubt to convict. This standard is higher than that employed in civil cases.

receiving stolen goods

same as *receiving stolen property.*

receiving stolen property

an offense that consists of accepting merchandise or other goods known to be or suspected of being stolen. Same as *receiving stolen goods.*

recidivism

engaging in unlawful behavior by those who previously were sentenced, treated, and released.

recidivist

one who relapses into criminal behavior.

Reckless, Walter C.

an American criminologist best known for developing *containment theory.* Educated at the University of Chicago, Reckless spent his academic career at Vanderbilt University and Ohio State University. See *Chicago School.*

record piracy

the illegal reproduction and distribution of copyrighted sound recordings. Also called bootlegging. The material may be reproduced in any of several formats, including vinyl, cassette tape, and digital compact discs. The advent of digital sound technology and computerized desktop publishing has made it possible for record pirates to manufacture albums whose sound recordings, artwork, and packaging are virtually indistinguishable from those of the legitimate product.

recreational killer

a serial killer whose primary motivation to kill is for pleasure.

recreational law enforcement

policing activities in areas where the public engages in recreational activities, such as hiking, snowmobiling, boating, or skiing.

red-handed

in the act of committing a crime.

red-light district

a section of a city in which there are a large number of houses of *prostitution*. Red-light districts take their name from the practice by brothel owners of using red lights to advertise their type of business.

reentry

the return and adjustment of an offender released from prison.

reentry court

a *specialty court* designed to facilitate the successful return of an offender from prison. Reentry courts can sentence an offender to a prison term with the provision that he or she will eventually be released under the supervision of the court. One advantage of reentry courts is the early identification of the offender's risks, needs, and assets so they can be addressed prior to sentencing, during confinement, and on release from prison.

referee

a legally trained court official who hears minor cases. Compare with *judge, magistrate*.

reformatory

a state correctional facility intended to rehabilitate youthful or otherwise nonserious offenders. Reformatories were first used in the 1800s.

Regional Information Sharing System (RISS)

a program consisting of six regional centers that share criminal intelligence information. RISS focuses on narcotics trafficking, violent crime, and gang activity. Membership in each RISS center ranges from 520 to more than 1,300 agencies. Federal agencies, including the *Federal Bureau of Investigation (FBI)*, the *Drug Enforcement Administration (DEA)*, the *Internal Revenue Service (IRS)*, and the *Bureau of Alcohol, Tobacco and Firearms (BATF)*, participate in RISS.

rehabilitation

a rationale for punishment that emphasizes correcting offender behavior through treatment. See *deterrence, incapacitation, retribution*.

reintegration

a rationale for punishment that emphasizes returning the offender to the community to facilitate links with family and employment. Compare with *rehabilitation.*

reintegrative shaming

shaming that has positive effects for the offender. Compare with *disintegrative shaming.* See *shaming penalties.*

relapse prevention

treatment services designed to prevent offenders from once again engaging in the maladaptive behavior that led to their legal problems. Relapse prevention is an especially important part of correctional programming for difficult-to-treat offenders, such as sex offenders and drug addicts.

release on recognizance (ROR)

the release of a charged offender without the posting of a monetary bond or surety. See *pretrial release.*

reparation

the act of making amends for a crime. Reparation can take many forms, including *restitution* and *community service.*

repeat offender

see *career criminal.*

reported crime

crimes reported to and therefore known to the police. For participating law enforcement agencies, reported crimes are included in the *Uniform Crime Reports (UCR)* or *National Incident Based Reporting System (NIBRS).* Compare with *hidden crime.*

resisting arrest

the refusal to comply with a law enforcement officer's attempt to detain. Also, the formal charge for such conduct. Resisting arrest as a charge usually accompanies one or more other charges that prompted the arrest.

restitution

money paid by an offender to victims to compensate them for injuries or loss of property. Restitution is generally combined with other legal consequences, such as *probation*, fines, and the payment of court costs.

restorative justice

a nonretributive justice philosophy that emphasizes restoration for the victim, the offender, and community rather than *retribution*.

restraining order

a court order requiring one individual to stay away from another. Restraining orders are commonly used by courts in cases of *domestic violence* to keep an abusive husband away from his wife. Compare with *civil protection order*.

retreatism

one of several modes of adaptation proposed by *Robert K. Merton* in his theory of *anomie*.

retribution

the rationale for punishment that expresses society's need for revenge. See *deterrence, incapacitation, rehabilitation*.

revenge

the motive or act of getting even for some real or perceived wrong. Revenge is expressed in the criminal law as *retribution*. See *vigilantism*.

revocation

the act or an instance of something being revoked, such as *probation* or *parole*.

revolver

a handgun in which the bullets are contained in a cylinder.

RICO

see *Racketeer Influenced and Corrupt Organizations (RICO)*.

rifling

the spiral grooves inside the barrels of firearms for the purpose of causing the bullet to spin and therefore stay on a straighter course. Rifling leaves marks on the bullet that *ballistics* experts can use to match the bullet to the firearm from which it was fired.

right to bear arms

an American right, grounded in the *Second Amendment* of the U.S. Constitution, allowing every citizen to own and possess firearms. Despite arguments that the right to bear arms is constitutionally guaranteed, some critics counter that the need for an armed militia in the 1700s is obviated by armed forces. See *National Rifle Association (NRA)*.

right to counsel

the right to have legal representation before a criminal court. Also, an area of study and policy focusing on this right. In the late 20th century, the *Office of Juvenile Justice and Delinquency Prevention (OJJDP)* made the right to counsel a priority area for study and funding.

rigor mortis

temporary muscular rigidity in a deceased person. Because rigor mortis is only temporary, it is used by homicide investigators to estimate the time of death.

riot

a civil disturbance, often resulting in violence. See *civil disobedience, public order crime*.

riot gun

a *shotgun* with a short barrel designed to intimidate or quell rioters. Riot guns, once considered unnecessarily offensive and consequently removed from police patrol cars, have been restored in many jurisdictions due to the heavy armament of some contemporary criminals.

risk assessment

the systematic process of determining the risk of reoffending.

risk assessment instrument

a pen-and-paper list of items designed to collect information about an offender for the purpose of determining future risk of reoffending.

risk-focused approach

the development of prevention and intervention programs for youth based on their risk of engaging in problem behaviors. Compare with *asset-focused approach*.

ritualism

in *Robert K. Merton's* theory of *anomie*, one of five modes of adaptation that an individual may adopt in the pursuit of social goals. Ritualism consists of going through the motions.

road rage

extreme anger and its associated violence by motorists.

roadblock

a barrier of vehicles created by law enforcement officers to stop fleeing offenders. See *Stop Stick*.

robber baron

a now dated term for a wealthy magnate who used power and influence to control industries through such tactics as manipulation of markets and intimidation of competitors.

robbery

the act of taking from another by force or threat of force. Robbery is often confused with *burglary*. See *armed robbery*.

rookie

a new, inexperienced police officer, generally in the first year of service.

routine activities approach

a theoretical perspective in criminology that suggests that crime occurs when three elements converge: (1) a suitable target, (2) a motivated offender, and (3) the absence of a capable guardian. The elimination of any one of the three elements theoretically makes the commission of a crime impossible. This theory on the genesis of crime has obvious implications for *crime prevention*.

Royal Canadian Mounted Police (RCMP)

the national police service of Canada. Administratively, the RCMP is an agency of the Minister of Canada. In addition to serving as federal Canadian police, the RCMP also provides contract services to territories, provinces, and municipalities.

runaway

a youth who runs away from home. Also, the charge officials level against one who runs away. See *status offender*.

rural crime

crime occurring in a rural area, particularly those offenses characteristic of such areas. An example of rural crime is the theft of gas from farm tanks.

Russian mafia

organized crime based in Russia or dominated by Russians. The Russian mafia became more powerful with the collapse of communism in the Soviet Union.

Rust Belt

any city or region in which the loss of local industry has harmed the economy, raised unemployment, and brought attendant problems such as urban decay and crime. When large factories that previously employed large numbers of people shut down, many employees are put out of work and can no longer contribute to the local economy. Businesses that previously depended on this commerce go out of business.

sabotage

intentional damage to plans or physical facilities, usually by an insider.

sadism

the derivation of pleasure from inflicting pain or suffering on others. See *sexual sadist*. Compare *masochism*.

sadomasochism

the combined practice of both *sadism* and *masochism*.

Safe Futures Initiative

an initiative designed to create a continuum of care in communities to prevent and control delinquency. Safe Futures is predicated on using community strengths, including youth.

safe house

a dwelling used by law enforcement or other criminal justice authorities to hide and protect witnesses or others who may be in danger.

Safety Town

a *crime prevention* program in which youth learn how to be safe in their communities. Topics covered in a typical Safety Town include how to cross the street, what to do in case of a fire, and what to do if approached by a stranger. Safety Towns, which emphasize all facets of a child's life, including home, school, and other environments, often end with graduation ceremonies for participating youth.

same-sex abuse

the physical, mental, or emotional abuse by a person toward another of the same gender.

sample

a group considered representative of a *population*.

San Francisco Project

a series of research projects undertaken in the late 1960s and early 1970s by the School of Criminology at the University of California, Berkeley to examine a number of practices related to *probation*. Among the studies in the San Francisco Project were inquiries into the *presentence investigation report*.

San Quentin

San Quentin State Prison, San Quentin, California. The oldest of California's prisons, San Quentin was established in 1852. It houses the state's *death row* for condemned men and the state's *gas chamber*.

sanction

a legal consequence for committing a crime.

sap

a heavy, leather-clap weapon used to strike a person. Compare with *truncheon*.

SARA model

acronym for scanning, analysis, response, and assessment. A method suggested by law professor Herman Goldstein to systematically document and evaluate problems in a community to address them. Scanning is the collection of data on problems in a community. Police make sense of these data through analysis, which leads them to a response to address identified problems. In the assessment phase, police officials evaluate their response to make any necessary adjustments. See *community policing, problem-oriented policing*.

satanic cult

a quasi-religious group whose practices and ceremonies center around the worship of Satan.

Satanist

one who worships Satan or otherwise practices black arts.

Saturday night special

an inexpensive, cheaply made, easily acquired handgun. Because of their low cost, Saturday night specials are said to be favored by those who want to engage in crime.

scaffold

a platform on which a condemned criminal is executed by hanging. Scaffolds typically have a trap door through which the condemned drops to his or her death.

Scared Straight

an inmate-run program that originated in the Rahway State Prison in New Jersey in the late 1970s and that was designed to acquaint youthful offenders with the grim realities of prison life. Scared Straight consisted of inmates, many of whom were serving life sentences for murder, who related graphic portrayals of physical and sexual assault and other forms of exploitation against young male inmates. The program was replicated throughout the United States by other inmate groups, and it was the subject of several television specials highlighting its methods. Evaluations of such programs showed that these had little effect on the subsequent criminal involvement of attendees.

school crime

crime committed in and around a school. School crime has always been a problem, but it has come to the fore due to a series of particularly devastating instances during the 1990s, including the *Columbine massacre*. School officials have been criticized for downplaying the extent and seriousness of school crime and for failing to keep or disclose crime-related statistics.

school safety

umbrella term that includes any program or movement intended to ensure the protection of students, teachers, and staff from intentional harm. School safety programs can include the use of school resource officers or metal detectors to prevent weapons from being brought into school.

scientific misconduct

illegal or unethical research or publication practices during the performance of scientific research. Scientific misconduct includes the fabrication or falsification of data, *plagiarism*, and the falsification of credentials by those who apply for scientific positions. Despite the fact that some forms of scientific misconduct are illegal, in most cases it is handled administratively.

Scotland Yard

the investigative division of the London metropolitan police. Scotland Yard has an international reputation for being a highly professional law enforcement organization.

search and seizure

the specialty in criminal law concerned with the ability of law enforcement to search criminal suspects and seize property as a result.

SEARCH, Inc.

an organization that since the early 1970s has promoted the development of criminal justice information systems in the United States. SEARCH, Inc., headquartered in Sacramento, California, receives much of its funding from the *Bureau of Justice Statistics (BJS)*.

search pattern

any of several systematic ways in which investigators comb crime scenes for *physical evidence*. Common search patterns include the spiral, wheel, grid, and zone methods, each named for the means by which the investigator covers the crime scene.

search warrant

a legal document issued by a judge or magistrate that permits law enforcement officials to enter and search specified environs, such as a home or business, for the purpose of finding evidence, apprehending a suspect, or for other legitimate purposes.

Second Amendment

an amendment to the U.S. Constitution that guarantees the citizens' *right to bear arms*. Second Amendment rights have been cited by the *National Rifle Association (NRA)* and other organizations. Opponents argue that when the Constitution's framers drafted its language, an armed citizenry was important to guarantee freedom. With the current well-developed armed forces, however, these arguments are less compelling.

secondary conflict

according to *culture conflict theory*, secondary conflict is that which develops in society as it becomes more heterogeneous. As groups differentiate and develop their own distinct values, they eventually clash with one another. Compare with *primary conflict*.

secondary deviance

in labeling theory, the deviance that results from the acceptance by an individual of his or her deviant identity. Compare with *primary deviance*. See *labeling theory*.

secondary prevention

the prevention of delinquency by those who are at risk or who have already engaged in delinquent acts. Compare with *primary prevention, tertiary prevention*.

secondary victim

one who indirectly suffers as the result of a crime, such as a relative or friend of a victim of a serious crime. Many secondary victims object to their experience being referred to as secondary because they have lost loved ones and consequently must live with grief and sorrow.

second-degree murder

murder that is not premeditated. Compare with *involuntary manslaughter, premeditation, voluntary manslaughter*.

secret police

law enforcement authorities of totalitarian governments who investigate and accuse, often without evidence. The KGB of the former Soviet Union is an example of secret police.

secretor

one who secretes blood type A, B, or AB antigens in body fluids. Criminal suspects who are secretors are more easily identified if body fluids are available at the crime scene or elsewhere.

securities fraud

any form of deceptive practice designed to illegally profit from the manipulation of stocks or other securities. See *Ivan Boesky, Michael Milken*.

security guard

a privately employed person whose job is to protect people and property. Sometimes referred to as rent-a-cops. Compare with *police*.

security threat group

a group of individuals in a correctional facility who collectively pose a threat to the safety and security of staff and other inmates. Examples of security threat groups are gangs, those affiliated with the *Aryan Brotherhood* or other white extremists, and black militants. Those affiliated with security threat groups are often segregated from one another by moving them to different institutions to minimize their ability to cause harm. Compare with *gang*.

sedition

actions or speech that could prompt rebellion against government or other authority.

selective incapacitation

the policy and practice of confining violent or chronic offenders based on statistical predictors, such as *prior record* and seriousness of the *instant offense*.

self-concept theory

any criminological theory that emphasizes the importance of a youth's self-concept as a cause of, or an insulator against, delinquency. See *containment theory*.

self-defense

acting in a way to save one's self from death or injury.

self-fulfilling prophecy

the tendency of a person to live up to the negative imputations made about him or her. Self-fulfilling prophecy is often used to describe delinquents who are labeled by the juvenile justice system. See *labeling perspective*.

self-incrimination

verbal or written statements that serve to cast suspicion on a person.

self-mutilation

the practice of cutting, piercing, burning, or otherwise disfiguring one's body. Self-mutilation is sometimes a characteristic of those who suffer from *borderline personality disorder*. Youth, particularly those having histories of abuse or neglect, may engage in such acts.

self-report study

a study in which survey respondents or interviewees are asked to reveal the nature and extent to which they have engaged in crime or delinquency. Self-report surveys, which can employ surveys or interviews, gained increasing popularity with the acknowledgment that traditional crime statistics such as the *Uniform Crime Reports* had severe limitations in their ability to accurately measure crime. Compare with *victimization survey*.

self-reported crime

crimes divulged by offenders in the course of a research study designed to measure, among other factors, the offenders' actual involvement in criminal activity. See *self-report study*.

Sellin, Thorsten

a prominent criminologist of the 20th century. Sellin was educated at the University of Pennsylvania, where he spent his entire academic career. He is perhaps best known for his monograph for the Social Science Research Council titled *Culture Conflict and Crime* (1938) and, with *Marvin E. Wolfgang* and Robert Figlio, *Delinquency in a Birth Cohort* (1972).

semiautomatic weapon

a firearm designed to discharge and rechamber a round with the single pull of the trigger. Semiautomatic weapons are particularly lethal due to the higher capacity of their magazines. See *automatic weapon, revolver*.

sentence

the legal consequence imposed by a court of law on a convicted offender. A sentence can consist of time in prison or jail, a fine, probation, restitution to the victim, other consequences, or any combination of these.

sentence bargaining

a form of *plea bargain* in which the negotiations between the prosecutor and defense attorney center around which sentence the prosecutor will recommend the judge impose. See *charge bargaining*.

sentence disparity

differences in sentences meted out to offenders convicted of like or similar offenses. Sentence disparities can be the result of legislative differences between jurisdictions or discretion.

sentencing

the phase of court processes at which the defendant is punished. In many felony cases, the presiding judge uses a *presentence investigation report* for assistance in arriving at a more just *sentence*. Sentencing options include *prison* or *jail* terms, *probation, fines,* or other alternatives.

sentencing circle

a form of restorative justice based on Native American tradition. See *circle sentencing*.

sentencing commission

an official body of appointed members charged with examining sentencing laws and practices and recommending changes.

sentencing discount

a reduction in an offender's sentence in return for a guilty plea. See *plea bargaining, sentence bargaining*.

sentencing guidelines

legislatively determined criteria for the imposition of criminal sentences. Sentencing guidelines restrict judicial *discretion*.

Sentencing Project, The

an organization that promotes the development of sentencing alternatives. Based in Washington, D.C., The Sentencing Project has published numerous reports, many pointing out inequitable and discriminatory practices in the criminal justice system. The Sentencing Project was a major force behind the *Campaign for an Effective Crime Policy*.

sentry

a lookout.

sequestering

the practice of isolating a jury from any outside influences, such as family, friends, and electronic and print media. Sequestering rests on the assumption that to render a just verdict in a particular case, jurors must not be biased by information from the outside. Depending on the nature and complexity of the criminal case, jurors may be sequestered for extended periods of time.

serial arson

the occurrence of two or more instances of arson by the same perpetrator. Like *serial murder*, serial arson is often sexually motivated. Compare with *fire setting*.

serial arsonist

one who engages in *serial arson*.

serial killer

one who commits *serial murder*. Also, serial murderer. Infamous serial killers include *Theodore Bundy, Jeffrey Dahmer, John Wayne Gacy*, and *Wayne Williams*.

serial murder

a series of homicides committed by one or more offenders with a cooling-off period between each one. Serial murder differs from most other forms of homicide in that it is sexually motivated. Compare with *mass murder, multiple murder, spree murder*. See *Theodore Bundy, Jeffrey Dahmer, John Wayne Gacy, Wayne Williams*.

serial murderer

same as *serial killer*. Compare with *mass murderer*.

seriousness scaling

the use of psychophysical scaling techniques to assess the perceived seriousness of criminal, delinquent, or deviant acts. Seriousness scaling in criminology was introduced by *Thorsten Sellin* and *Marvin E. Wolfgang* in *The Measurement of Delinquency* (1964). The two major methods of seriousness scaling are category scaling and magnitude estimation scaling.

serology

the study of blood.

serotonin

a compound in the blood that also acts as a neurotransmitter. Some believe that serotonin levels influence criminal behavior. See *biocriminology*.

Serpico, Frank

a former New York City police officer who, in the early 1970s, exposed widespread corruption within the New York City Police Department. Serpico met considerable hostility from other officers, many of whom were involved in the corruption. His allegations led to the formation of the *Knapp Commission* and resulted in the conviction of numerous police officials at all levels.

service bailiff

a bailiff employed by a court who serves summons, subpoenas, and other legal documents that must be delivered in person.

service revolver

see *service weapon*.

service weapon

the weapon, most often a handgun, that a law enforcement officer carries while on the job. Compare with *off-duty weapon*.

severity

the punitiveness of a criminal sanction. Severity is one of several characteristics of criminal sentences focused on by adherents of the *classical school of criminology*. Compare with *celerity*, *certainty*.

sex offender

one who engages in sex-related crimes, such as rape or child molestation.

sex offender registration

the requirement by law that convicted sex offenders must register with local law enforcement agencies. Sex offender registration became popular in the 1990s in the wake of several highly publicized sex crimes by those with histories of such offense. See *Megan's law*.

sex slave

a human captive, most often female, who is forced to engage in sexual relations with the captor.

sex trade

the selling and transportation of females for use in *prostitution*.

sexploitation

term used to describe the commercial exploitation of sex through the production and distribution of sexually explicit materials. See *pornography*.

sexual deviance

any deviant sexual practice.

sexual homicide

a homicide in which the perpetrator's motive is primarily sexual in nature.

sexual predator

an offender who chronically commits sexual offenses. The term sexual predator has also been used by legislators to designate persons at whom special legislation is aimed.

sexual psychopath

a sex offender considered predatory and beyond rehabilitation.

sexual sadist

a sexual offender who derives pleasure from inflicting pain on his or her victims.

shackles

metal restraints that close around the ankles or wrists, linked to chains. Compare with *handcuffs, leg irons*. See *belly chain*.

shakedown

the systematic, often unannounced, search for contraband in jails or prisons. Shakedowns routinely yield homemade weapons, drugs, and other contraband.

shaming penalties

consequences for crime where the offender is subjected to the community's disapproval. Shame penalties have their beginnings in Australia and New Zealand. See *disintegrative shaming, integrative shaming*.

shank

slang term for a homemade knife. Shanks, which can be fashioned from a variety of hard materials such as spoons and bedsprings, are common in prisons and other correctional facilities. Periodic shakedowns in prisons invariably turn up a variety of contraband, including a number of shanks. Inmates are adept at hiding shanks. Also referred to as *shiv.*

Shaw, Clifford

an American criminologist who, with Henry D. McKay, developed the *ecological school of criminology*. Shaw received graduate training at the University of Chicago, where he was exposed to the work of Ernest Burgess and Robert Park. See *Chicago Area Project, Chicago school, Henry D. McKay.*

Sheldon, William

(1898-1977) a medically trained professor of psychology who developed a *typology* relating body types to temperament. See *constitutional theory, ectomorph, endomorph, mesomorph.*

shell games

games of chance in which the player must guess the location of a pea under one of several shells. Unwary players do not realize that shell games are rigged and therefore impossible to win.

shelters

places where victims of domestic violence can temporarily live in safety. Shelters, most often operated by nonprofit organizations, do not disclose their locations to the public for purposes of safety and security.

sheriff

an elected official of a county whose duties include enforcing laws and serving legal notices. In the United States, sheriffs typically provide law enforcement services in unincorporated areas of a county. They also operate county jails, in which offenders serve sentences and await conveyance to state prisons. In some areas, sheriffs also secure the courts and deputies serve as bailiffs.

Sherman Antitrust Act

a law passed in 1890 to regulate corporate behavior. The Sherman Antitrust Act prohibits the development of monopolies and provides penalties for *price-fixing.*

shiv

same as *shank.*

shock incarceration

a sentence in which the offender serves a brief period of confinement in prison followed by release on probation or parole. The theory underlying shock incarceration is that exposure to the harsh realities of prison life will shock the offender into remaining crime-free. See *shock parole, shock probation.*

shock parole

parole following a brief period of confinement in prison designed to acquaint the offender with the realities of prison life. Offenders receiving shock parole typically spend more time in prison than those participating in *shock probation.*

shock probation

probation following a brief period of confinement in prison designed to "shock" the offender with the realities of prison life. Generally, the sentencing judge imposes a term of incarceration. After a designated period of time, the inmate may then apply to the sentencing court to have a shock probation hearing. The inmate, either present or in absentia, is either granted or denied probation. If shock probation is granted, the inmate is released from prison under probation supervision for a specified period of time under certain conditions.

shock treatment

a medical procedure in which a mental patient is given an electroconvulsive shock to restore him or her to improved mental health. Also referred to as electroconvulsive shock therapy.

shoplifting

theft in a retail store by concealing merchandise. See *booster girdle.*

shotgun

a *long gun* with a smooth bore that fires numerous small pellets. At short ranges, shotguns can cause devastating injuries. Compare with *rifle.*

shunning

the practice of avoiding those who have violated certain norms. Shunning is most closely associated with Amish culture. See *shaming.*

sight and sound separation

the requirement that juveniles held in adult jails or lockups must be removed in such a way that the adults and juveniles cannot have contact. Sight and sound separation is based on the notion that any contact with adult inmates is detrimental to confined juveniles.

signature

a distinguishing trademark of a serial killer or other offender evident at the crime scene. Examples of signatures include types of binding, the use of unusual knots, peculiar injuries, or the position in which the body is left by the offender.

silencer

a device, generally affixed to the muzzle end of a firearm, designed to suppress sound. Even though silencers are illegal to possess, the constituent parts necessary to construct them can occasionally be found for sale at gun shows. Also referred to as suppressor.

Silkwood, Karen

a woman who in 1974 took on the nuclear plant in which she worked for its disregard of health and safety standards. After gathering incriminating documents, she was en route to a meeting when she was killed in an auto accident. The documents disappeared from the accident scene. The nuclear plant, Kerr-McGee, subsequently was found guilty of contaminating Silkwood and was ordered to pay her estate $10.5 million in damages. See *corporate crime*.

Simpson, O. J.

football and movie star who in 1994 was accused of killing his ex-wife Nicole Simpson Brown and her friend Ronald Goldman. Simpson's initial flee from authorities and subsequent trial captured the attention of the world, with the combination of high-priced defense attorneys known as the *Dream Team* and alleged mistakes by law enforcement authorities. Although Simpson was acquitted of the murder charges, he was sued in civil court and lost. Many still believe he is guilty, especially in light of compelling *DNA testing* evidence.

Sing Sing

an infamous New York state prison on the banks of the Hudson River near Ossining, New York.

situational crime prevention

the prevention of crime by recognizing that offenders make rational choices to engage in crime and by removing or reducing opportunities for offending. An example of situational crime prevention is the placement of jewelry in locked display cases that only store clerks can open.

sketch artist

an artist who sketches renditions of suspects based on the recollections of witnesses. See *Identi kit.*

skimming

to hide profits, often to avoid the payment of taxes.

skinhead

a member of a white gang who wears closely cropped hair and tattoos and uses violence against members of certain racial and ethnic minorities. Skinheads subscribe to neo-Nazi beliefs.

slander

spoken statements that negatively affect the name or reputation of another. Compare with *libel.*

slap jack

a leather, lead-filled device used for hitting.

slavery

the practice of keeping humans in bondage against their will. See *white slavery.*

smack

slang for heroin.

small arms

firearms capable of being carried. These include *handguns* and *long guns.*

Smith, Susan

a woman who murdered her two children by rolling her car into a lake in South Carolina. Smith, who claimed a carjacker had taken her car with the children still in it, made tearful pleas on television for the safe return of her children. She was convicted of two counts of first-degree murder and was sentenced to life imprisonment.

Smith & Wesson

since the 1800s, a major manufacturer of firearms, especially handguns. Smith & Wesson, along with other gun makers, was the subject of lawsuits in the 1990s for manufacturing dangerous consumer products.

smuggling

the surreptitious transfer of goods, often for the purpose of avoiding the payment of tariffs. See *U.S. Customs Bureau.*

snitch

compare with *booster.*

snuff film

a film in which a person is actually killed on camera. For many years, snuff films were the stuff of urban legends in that people referred to them but no one had ever actually seen one.

social bond theory

the criminological notion that the strength of a person's bonds to society affects the likelihood of becoming involved in delinquency and crime. Social bond theory is most closely associated with criminologist Travis Hirschi. See *control theory.*

social capital

personal assets that increase the likelihood of positive social adjustment and, conversely, help insulate the individual from *criminogenic factors.*

social class

the social and economic stratum to which a person belongs.

social contract

the unwritten agreement between self-interested individuals and the government in which they consent to restrict their pursuit of self-interest in pursuit of peace. In return, the government agrees to protect the individuals' rights and freedom.

social control

both formal or informal means of ensuring compliance with laws, rules, and norms.

social control theory

any one of several perspectives in criminology that emphasize crime as a consequence of uncontrolled, hedonistic individuals. See *containment theory*.

social defense

a nonviolent, nonmilitary response to aggression. *Social justice,* which favors demonstrations and boycotts over the use of weapons and force, includes both domestic and foreign aggression.

social disorganization

the problems besetting communities facing large-scale immigration, the loss of industry, and high unemployment. Socially disorganized communities typically experience higher rates of *crime* and *delinquency*.

social ecology

the field concerned with the relationship between humans and their environment.

social justice

equity and fairness in all spheres of life, including but not limited to economic opportunity, gender, the freedom of expression, concern over natural resources, and human rights.

social learning theory

also *learning theory*.

social norms

unwritten rules by which people in a society abide.

social pathology

behavior in society that has negative consequences, such as *crime* and *delinquency*.

social therapeutic institutions

in Europe, a facility of convicted offenders designed to emphasize treatment over punishment. Incorporated into penal codes, social therapeutic institutions were criticized by conservatives for not being retributive enough and by liberals for perpetuating the medical model of corrections. See *therapeutic community*.

Society for Laws Against Molesters (SLAM)

an organization committed to promoting tougher legislation to punish child molesters.

sociopath

an individual who has no conscience, does not learn from experience, and has little regard for the rights or feelings of others. See *antisocial personality disorder, psychopath.*

sodomy

sexual intercourse with a member of the same sex or with an animal. Sodomy also refers to anal or oral sex between humans.

software piracy

the illegal acquisition and reproduction or use of licensed computer software.

solicitor

in Britain, a member of the legal profession qualified to advise clients and provide instruction to barristers. Compare with *barrister, counselor.*

solitary confinement

see *hole.*

somatotypes

body types that are believed to have a relationship to an individual's personality. See *ectomorph, endomorph, mesomorph, William Sheldon.*

Son of Sam

the pseudonym used by *David Berkowitz,* the infamous .44-caliber killer who, in the 1970s, shot and killed men and women he found at lovers' lanes and other isolated locations.

Son of Sam laws

any state or federal laws designed to prohibit convicted criminals from profiting from their crimes by way of lucrative book or movie deals or by other means of commercial exploitation. These laws take their name from the *Son of Sam* killings by *David Berkowitz,* who attempted to sell his story to a major publisher. New York was the first state to enact such a law. These laws have been controversial because they conflict with the First Amendment of the *Bill of Rights.*

sororicide
the killing of a sister by a sibling.

Sourcebook of Criminal Justice Statistics
since 1973, an annual compilation of a wide variety of crime- and justice-related statistics. The *Sourcebook of Criminal Justice Statistics*, which is also available on CD-ROM, includes data on arrests, reported offenses, victimization, citizen surveys, and correctional populations. Funded by the *Bureau of Justice Statistics (BJS)*, the *Sourcebook of Criminal Justice Statistics* is prepared by the Hindelang Criminal Justice Research Center at the State University of New York at Albany.

Southern Police Institute
a law enforcement education and training center affiliated with the Department of Justice Administration at the University of Louisville, Kentucky.

Southern Poverty Law Center
a legal organization located in Montgomery, Alabama, that tracks white supremacist and other hate groups and engages in litigation against them. The Southern Poverty Law Center is a nonprofit organization and is funded by contributions.

souvenirs
in a *serial murder*, an object taken by the killer from the victim as a memento of the crime. Souvenirs can be anything from the victim's personal possessions to body parts. The identification of souvenirs can help law enforcement officials make linkages among related homicides.

specialty court
a court designed to address a specific social issue or offender problem, such as domestic violence, drugs, guns, and mental health. Proponents argue that specialty courts are an improvement over traditional courts in that they better meet the needs of the offender and society. Those opposed to specialty courts believe that traditional courts can adequately handle a variety of diverse cases. See *domestic violence court, drug court, gun court, mental health court*.

specific deterrence
the type of deterrence directed toward individuals who have already engaged in crime. Locking up an offender is an example of

specific deterrence. Also referred to as *individual deterrence.* Compare with *general deterrence.*

specification

a circumstance of a crime that carries an additional mandatory term of confinement for those convicted. An example of a specification is the use of a firearm during the commission of a robbery. Aside from the penalty that the robbery carries, the law also specifies that the use of a gun results in an add-on of additional years.

Speck, Richard

(1941-1991) an infamous mass murderer convicted of stabbing and strangling eight student nurses in Chicago in 1966. Although convicted and sentenced to death, Speck was spared execution by the 1972 prohibition against the death penalty. Years after his conviction, Speck again made news by his participation in videotaped homosexual acts while in prison. See *mass murder.*

spectator violence

violence perpetrated by those who attend sports events. Spectator violence by hooligans at European soccer games has resulted in numerous deaths and injuries. Much of the problem can be attributed to the excessive consumption of *alcohol.* See *hooliganism.*

speedy trial

the right of a defendant to have the pending criminal case brought to trial within a specified period of time. The defendant may waive this right, permitting the criminal justice system to take longer to process the case.

Spike Strip

See *Stop Stick.*

split sentence

a sentence given in a criminal court that consists of a term or confinement coupled with conditional release such as probation. Split sentences combine the *incapacitation* implied in imprisonment with the opportunities presented by *community-based sanctions.*

sport violence

violence that often accompanies certain forms of athletics. Boxing is perhaps the most controversial sport because violence is not a by-product but an expected part of the action. Certain groups of medical professionals, including the American Medical Association, have spoken out against sport violence. See *hooliganism*.

spousal abuse

the psychological or physical mistreatment of one spouse by the other. Compare with *domestic violence*.

spree murder

a series of homicides committed within a short period of time, sometimes in conjunction with other felonies, such as robbery or sexual assaults, without any intervening cooling-off period. A notorious example of spree murders are those of *Charles Starkweather*, who, in 1956, killed 10 people in 8 days. Spree murder is a form of *multiple murder*. Compare with *mass murder*, *serial murder*.

spuriousness

an apparent but false relationship between two variables. The classic example of a spurious relationship regarding crime is that between ice cream sales and crime: As ice cream sales increase, so does crime. This might lead some to conclude that ice cream sales cause crime. The relationship is spurious, however, because both ice cream sales and crime are related to a third variable—hot weather.

spy

a person who secretly collects and reports information on the activities, movements, and so on of an enemy or competitor. See *espionage, Walker spy ring*.

staging a crime scene

the arrangement of a corpse, weapon, or other evidence by a police officer to mislead investigators about how a crime took place. Investigators have become adept at identifying a staged *crime scene*.

stalking

the following of one person by another, often undetected, for the purpose of harassment or with the intent to do bodily harm. Stalking gained national attention in the 1980s and 1990s when several Hollywood celebrities were stalked. Many states have enacted legislation designed to punish stalkers and protect their victims.

Starkweather, Charles

(1938-1959) a *spree killer* who, with teenaged girlfriend Carol Ann Fugate, went on an infamous murderous rampage throughout the Midwest. He was executed in 1959.

starring

the star-shaped rupture of the human flesh caused by the pressure of exploding gases during a contact gunshot wound. Also referred to as a *stellate*. See *contact wound*.

state administering agency

in U.S. Department of Justice nomenclature, a state agency that administers federal formula or block grant programs, such as those of the *Bureau of Justice Assistance (BJA)* or the *Office of Juvenile Justice and Delinquency Prevention (OJJDP)*. Many but not all state administering agencies also serve as state planning agencies. See *state planning agency*.

State Justice Institute (SJI)

a private, nonprofit organization created by Congress in 1984 to foster improvements in state courts. The SJI periodically awards competitive grants to organizations and governmental agencies to address court-related problems or to undertake innovative practices. Projects sponsored by SJI include those related to training for judges and court personnel, substance abuse, and the application of technology. The SJI is headquartered in Alexandria, Virginia.

state planning agency

a unit of state government officially charged with overseeing criminal justice planning and policy development throughout the state. Many state planning agencies were created in the late 1960s to administer the federal funds made available through the *Law Enforcement Assistance Administration (LEAA)*. Some state planning agencies are also *state administering agencies*. See *National Criminal Justice Association (NCJA)*.

state police

a law enforcement agency that has broad legal authority to investigate violations of criminal and traffic laws anywhere within the state. Home-rule states do not have state police but instead rely on local law enforcement authorities such as sheriffs for enforcement.

state prison

a correctional facility operated by state authorities for the confinement of convicted felons. Compare with *federal prison*.

stationary killer

a *serial killer* who commits his or her crimes within a small geographical area. Compare with *nomadic killer*.

stationhouse release

the release of an offender immediately after arrest without the filing of formal charges. Stationhouse release is an example of law enforcement's use of *discretion*. Compare with *diversion*.

statistical analysis centers (SACs)

an organization unit in most U.S. states and territories formed for the purpose of collecting, analyzing, maintaining, and disseminating various criminal justice data. In addition to serving as repositories for crime and justice data, many SACs undertake cross-sectional studies of such topics as inmate suicide and police misconduct. See *cross-sectional study*.

statistical prediction

the use of statistical variables to predict the likelihood of some outcome. In criminological research, the outcome of interest might be offender *recidivism*. Compare with *clinical prediction*.

status offender

a minor whose offense would not be illegal if he or she were an adult. Status offenders include those charged with *runaway* and *incorrigibility*. The federal *Office of Juvenile Justice and Delinquency Prevention (OJJDP)* has made a commitment to force states not to detain status offenders in secure facilities, especially in close proximity to adult offenders.

status offense

an offense committed by a juvenile that would not be illegal if the person were an adult. Examples of status offenses are *incorrigibility*, running away from home, and *truancy*.

statutory rape

sexual intercourse with a person under the age of consent. Statutory rape, although defined as illegal, often involves sex between consenting individuals. In most jurisdictions, the age difference between the offender and victim is significant in defining the act as statutory rape.

stellate

a star-shaped wound caused by a contact gunshot wound to the head or body. Gases expand under the skin, causing a ragged wound resembling a star or cross. See *contact wound*.

stick-up boy

a small-time armed robber. See *robbery*.

stigma

an undesirable characteristic or reputation that follows youth, those with mental illness, or others processed by agents of *social control*. Stigma can reduce future opportunities for the person labeled. See *labeling perspective, radical nonintervention*.

stimulant

a drug that creates a sense of euphoria in the user. Other common effects of stimulants include sleeplessness and loss of appetite. *Amphetamines* are an example of a stimulant.

sting operation

an organized, secret effort by law enforcement authorities to investigate and arrest those suspected of engaging in professional or organized criminal activity.

stippling

same as *tattooing*.

Stockholm syndrome

a phenomenon in which a hostage begins to identify with and grow sympathetic to the captor and antagonistic toward the authorities. The syndrome derives its name after the city where, in 1973, a female bank robbery hostage became emotionally attached to one of her captors. One of the most famous cases of the Stockholm syndrome was that of newspaper heiress Patty Hearst, who was kidnapped by members of the Symbionese Liberation Army in 1974.

stocks

a wooden structure used in colonial times to publicly punish minor offenses. Compare with *pillory*.

stoner gangs

gangs of white youth who identify with heavy metal music, punk identities, and sometimes satanic cults. Some stoner gangs have affiliated with white extremists. The term stoner derives from a person being stoned on drugs. See *White Aryan Resistance*.

STOP

see *Stop Turning Out Prisoners*.

Stop Stick

a long, triangular box filled with sharp steel spikes designed to stop the motor vehicle of a fleeing suspect by puncturing the tires. Law enforcement officers lie in wait for the suspect vehicle and then pull a rope, drawing the Stop Stick into the vehicle's path. Once the tires pass over the Stop Stick, hollow spikes penetrate and insert in the tire, causing deflation and allowing pursuing officers to more easily apprehend the suspect. Also known as a *Spike Strip*.

Stop Turning Out Prisoners (STOP)

a movement that originated in Florida devoted to stemming the early release of violent prison inmates. STOP proposed legislation and undertakes publicity campaigns.

strain theory

any theoretical perspective in criminology that emphasizes the strain among individuals, particularly the disconnect between socially desirable goals and the available means to attain them. See *anomie theory*, *social disorganization*.

stranger abduction

the *abduction* of an individual by someone unknown to the individual. Stranger abductions, although capable of creating fear in parents for their children, comprise a relatively small proportion of all abductions.

strangling

the act of squeezing the throat of a person, thereby cutting off the air supply and eventually resulting in serious injury or death.

street crime

conventional, unsophisticated crime, such as robbery and theft. Street crime, which does not require special skill or education to commit, comprises the majority of crime in the United States. Compare with *upperworld crime.*

strict liability

a theory of close adherence to the letter of the law, even in the face of ignorance that a violation has occurred. According to strict liability, a person who unknowingly buys stolen goods is criminally liable even though he or she was not aware that the merchandise was "hot." Compare with *vicarious liability.*

strip search

the search of an accused suspect or convicted offender that involves the removal of clothes and the search for weapons and other contraband over the entire person, often including body cavities. Strip searches are necessary because some criminals are adept at hiding contraband.

structural Marxism

a Marxist approach in which the existence of capitalism fails to explain the role and control of technology in promoting stratification and change in society. Compare with *instrumental Marxism.*

structural theory

any theory in criminology that emphasizes the role of immutable structures in society in the genesis and transmission of crime and delinquency.

student threats

threats made by students against teachers, other students, or the school. See *school crime.*

stun gun

an electrical hand-held device that permits the user to render another person temporarily immobile through the application of a nonfatal, high-voltage electrical charge. See *Taser.*

sub rosa indictment

a secret indictment. Sub rosa indictments are often used when officials fear that any publicity will compromise their ability to successfully apprehend and prosecute the party in question.

subculture

a smaller cultural group with beliefs, *norms*, practices, and rituals that are different from, and sometimes at odds with, those of the larger culture. Examples of subcultures are *gangs* and the *homeless*.

subculture of violence

A term coined by criminologists *Marvin E. Wolfgang* and Franco Ferracuti to describe a subculture in society whose norms support violence as an acceptable way of life. An example of a subculture of violence is that found in the southern United States, where affronts to one's honor are typically met with a violent response. Gangs, also subcultures of violence, operate under different norms from those of larger society.

subculture theory

any criminological theory that focuses on the distinct characteristics or criminogenic effects of a subculture.

Substance Abuse and Mental Health Services Administration (SAMHSA)

a large division of the U.S. Department of Health and Human Services responsible for promoting programs and providing funding for substance abuse and mental health issues.

substantial capacity test

a test of insanity that asks whether the accused had the substantial capacity to appreciate the wrongfulness of the conduct in question and, if so, to control that behavior. See *Durham rule*.

subterranean values

the values of deviant subcultures such as *delinquents*.

suffocation

a condition caused by the interruption of the flow of air for breathing. Suffocation is a common cause of death in homicides, especially those of infants.

suicide

the intentional taking of one's own life. Suicide is actually a form of *homicide* and is investigated as such.

suicide by cop

the intentional effort by a person to force a police officer to shoot. Some suspects refuse to surrender their weapons and even charge the police, leaving officers no alternative but to use *deadly force*. With the successful development and adoption of *nonlethal weapons*, there should be fewer such incidents.

superior court

a court in which felony cases are heard.

supermax prison

a maximum-security prison or unit designed to house the most dangerous convicted prisoners. Because supermax prisons segregate offenders and keep them in isolation, it has been argued by some that they represent cruel and unusual punishment.

superpredator

term used to describe youthful offenders capable of greater amounts of more serious criminal behavior than their predecessors. Some criminologists predicted that in the early 21st century superpredators would proliferate, creating a wave of violent crime.

Supplementary Homicide Reports (SHRs)

an adjunct on criminal homicide reported routinely by local law enforcement agencies as part of their participation in the *Uniform Crime Reports* program. Data collected through SHRs include the offender's age, gender, and race; the victim's age, gender, and race; the circumstances of the offense; the offender-victim relationship; and the type of weapon. SHRs have largely been replaced by the *National Incident Based Reporting System (NIBRS)*.

supply reduction

a drug-control policy that emphasizes lessening the desire in current and potential users. Compare with *demand reduction*.

supreme court

in courts of appeals, the highest court to which an appeal can ascend. Rulings of the U.S. Supreme Court are final.

surety

one who assumes responsibility for the appearance in court of another.

surveillance

close watch over something or someone, such as a criminal suspect. Known informally as stakeout, surveillance by law enforcement involves the surreptitious scrutiny of suspects or locations.

surveillance equipment

mechanical or electronic devices designed and used to monitor others. Examples of such equipment are fiber-optic cameras and small listening devices also referred to as bugs.

survival analysis

a statistical technique used to measure the time between the onset of a disease and a terminal outcome. Survival analysis, originally developed to study the survival of patients with diseases such as cancer, has been employed by criminologists to measure how long offenders remain crime-free after they have undergone various forms of treatment.

survivalist

a radical, often heavily armed individual who fears takeover by foreign or domestic government. Survivalists stockpile provisions in the event they have to retreat from mainstream society and defend their freedom and beliefs.

survivor

term used to describe the living victims of crime, especially violent offenses such as domestic violence, that often result in serious injury or death. The meaning of the term has been broadened to include the family and close friends of the victims of violent crimes, including homicide, who continue to suffer long after the commission of the crimes. See *secondary victim*.

suspect

A person believed by authorities to be responsible for a crime. Compare with *accused, arrestee*.

suspense novel

a book-length piece of fiction in which the main character is placed in jeopardy, creating suspense on the part of the reader. Compare with *mystery*.

Sutherland, Edwin H.

(1880-1951) a leading figure in 20th-century American criminology, considered by many to be the father of criminology. Sutherland received graduate training at the University of Chicago and taught at several universities before becoming a faculty member at Indiana University, where he remained for the rest of his career. His most notable contributions are *The Professional Thief* (1937), *White-Collar Crime* (1949), and *Principles of Criminology* (1924), a textbook coauthored with *Donald Cressey* for which 12 editions were printed over 40 years.

SWAT

Special Weapons and Tactics. A specialized squad of law enforcement officers whose job is to handle high-risk operations, including hostage negotiations, *high-risk entry*, and violent felonies in progress. SWAT officers are typically trained in rapelling and scaling as well as in the use of a variety of weapons.

swindling

the cheating of a person out of money or possessions.

switchblade

a spring-loaded knife designed to open quickly with the push of a button. Switchblades were made illegal in the United States by federal law.

symbolic interactionism

a social psychological perspective that emphasizes subjective impressions and interpretations in human interaction. Symbolic interaction, which derived from the work of early social psychologists Charles Horton Cooley and George Herbert Mead, laid the theoretical groundwork for the *labeling perspective* in criminology.

syndicate

a large criminal organization. The word syndicate is frequently used for *organized crime*.

systematic check forger

forgers who purposely engage in forgery as a business. Compare with *naive check forger*.

tagging

the practice by *gang* members or other youth of spray painting *graffiti*.

Tailhook scandal

an incident involving U.S. Navy personnel in which male officers assaulted 26 women (including other Navy personnel) at a convention of the Tailhook Association in Las Vegas in 1992. Although the officers' behavior was cast as sexual harassment, it easily could have been charged as sexual assault.

tailing

the practice of following another person, generally in a motor vehicle, to maintain visual contact or to determine the person's destination. Implicit in tailing is that the *surveillance* is not detected by those being tailed.

tampering with evidence

the illegal removal, destruction, or alteration of *evidence* in a criminal case.

Tarde, Gabriel

(1843-1904) a French magistrate and sociologist who developed the forerunner of learning theory in criminology. Tarde, who rejected the work of *Cesare Lombroso* and other constitutional theorists of the day, argued that people become criminals by imitating the behavior of others. His work influenced a number of later criminologists, including *Edwin H. Sutherland*. See *learning theory*, *social process theory*.

target removal

efforts to eliminate the potential object of a criminal's intentions. Target removal, which is consistent with the *routine activities approach* in criminology, is based on the notion that criminals cannot victimize what is unavailable. An example of target removal is the use of *The Club* to prevent auto theft. See *crime prevention*.

TASC

see *Treatment Alternatives to Street Crime*.

Taser

an electronic weapon that works by discharging high-voltage electrodes attached to long wires that, when they penetrate human flesh, render the individual temporarily immobile. Originally designed for self-protection by law enforcement, security, and other well-meaning, innocent persons, it has been used by criminals for committing assault. Compare with *stun gun.*

tattooing

tattoo-like alterations to human skin caused by grains of gunpowder being driven into the skin. The presence of tattooing can reveal the distance between a shooter and the person shot.

tax evasion

the purposeful attempt to not pay taxes to which the government is entitled.

tax haven

a location, most often a country or possession, where income tax is low or nonexistent. Tax havens are used by those who want to avoid paying taxes in their country of residence.

tearoom

term used for public bathrooms, such as those found in parks or roadside rest areas, where homosexual activity takes place.

tearoom trade

impersonal homosexual activity in public restrooms or other public places.

technical violation

an infraction of the rules of *probation* or *parole*. Technical violations, which are not new criminal charges, can include such infractions as moving without permission or associating with a *codefendant*. See *parole violator*.

teen court

a quasi-judicial process operated by teenaged youth for the purpose of settling disputes and learning about the justice system. See *specialty court*.

telemarketing fraud

the illegal solicitation of donations or business by telephone that results in undervalued or no merchandise for the purchaser. Such schemes are sometimes used simply to get the credit card numbers of unwary customers.

territorial killer

a serial killer who commits crimes within an identifiable, circumscribed geographical area.

terrorism

the advocating or use of violence to coerce government action. Terrorism differs from other forms of crime in that it is most often motivated by political or religious considerations. Examples of terrorism include the *hijacking* and crashing of commercial jets into the World Trade Center and the Pentagon on September 11, 2001. Terrorism differs from other forms of crime in its motivation and *modus operandi*. See *Attack on America, domestic terrorism, international terrorism*.

terrorist

one who engages in terrorism. *Timothy McVeigh*, who was executed in 2001, and *Osama bin Laden* are notorious terrorists of the late 20th and early 21st centuries.

tertiary prevention

the prevention of undesirable traits or behavior in those who have already been afflicted with a condition or disease. The tertiary prevention of crime is more the control of future offending. The tertiary prevention of violent crime, for example, might consist of *incapacitation* combined with intensive treatment. Compare with *primary prevention, secondary prevention*.

theft

the intentional taking of property. See *grand theft, petty theft, theft by deception*.

theft by deception

illegally gaining the property of another by using lies or other forms of deception.

therapeutic community

a treatment philosophy and related practice based on the segregation of clients. Research has shown that therapeutic communities have value in the treatment of substance abusers.

therapeutic jurisprudence

an approach to the use of law as a potential therapeutic agent. Therapeutic justice rests on the assumption that the law has the power to effect dramatic social consequences, and that these consequences should be considered and studied. Of special interest in therapeutic jurisprudence is law's impact on the psychological and emotional well-being of individuals.

thermic law of crime

Adolphe Quetelet's conception that crimes against persons tend to occur in warmer climates, whereas property crimes are more likely to occur in colder climates. Empirical support exists for the thermic law of crime.

thief

a person who commits *theft*.

three strikes and you're out

a penal philosophy policy expressed in state laws that imprisons for life offenders who attain their third conviction. Three strikes and you're out, which has its beginnings in California, has been controversial because some offenders can in fact strike out for a relatively minor offense such as shoplifting. Consequently, it may not further the ends of justice. It has also been criticized because there is little reason to believe such a policy will have an appreciable effect on the amount of crime.

throwaway

a weapon left by a law enforcement officer at the scene of the shooting of an unarmed suspect. The throwaway is intended to make the officer's shooting of the suspect appear justified.

ticket of leave

in colonial Australia, the official permission given to an offender for the purpose of establishing a residence and finding employment prior to the expiration of the sentence. The ticket of leave is said to be the precursor of modern-day *parole*.

time series analysis

a statistical technique used to forecast future trends in crime rates, correctional populations, or other criminal justice applications. Time series analysis analyzes trends over periods of time. Compare with *ARIMA*.

tipping point

the point in an epidemic at which the disease spreads rapidly throughout a *population*. Concerning crime, the tipping point occurs when crime in an area increases to a certain level at which it takes on epidemic proportions.

tire marks

surface impressions made by a vehicle's tires. When tire marks are found at a crime scene, investigators can use a combination of photography and casting to determine the make and model of a particular tire through laboratory analysis. Such castings can later be compared to suspect vehicles to determine the vehicle's involvement in a crime.

toe tag

An identification tag affixed to the toe of a deceased person by a coroner or other official.

token economy

a means of exchange used by residents in a *therapeutic community*.

Tongs

Chinese organizations that engage in a variety of criminal enterprises, including extortion.

tool-mark identification

the analysis of scratches and other marks found in metal and other hard materials at crime scenes for the purpose of making linkages between the implements and the crime in question.

torture

the infliction of pain on another, often for the purpose of forcing the tortured person to reveal information. Torture throughout the world is closely monitored by *Human Rights Watch*.

total institution

a setting in which individuals are isolated from larger society. Total institutions are bureaucratic, rigid, and closely control the lives of their charges. A prison is an example of a total institution.

toxic criminals

a term applied to corporations that create environmental disasters through the dumping or otherwise irresponsible handling of hazardous waste. See *corporate crime*.

toxicology

the study of poisons and other toxic substances.

trace evidence

physical evidence left behind as the result of a crime.

training school

a residential school in which delinquents receive vocational training.

trait theory

any criminological theory concerned with the traits of offenders.

trajectory

the path of a bullet or other projectile. See *ballistics*.

transnational crime

organized crime, corruption, and other forms of criminal behavior whose influence spans the globe.

transportation

the movement of convicted criminals from one geographical location to another, generally more remote location. Nonindegenous Australia was first populated with convicts transported from England.

treason

the act of violating one's allegiance to a country by attempting to overthrow or otherwise subvert the government. Treason is an extremely serious crime, often punishable by death. See *espionage, Walker spy ring.*

Treatment Alternatives to Street Crime (TASC)

a national program designed to provide substance abuse treatment for offenders involved in the criminal justice system.

trespassing

the unauthorized presence of a person on the property of another that is prohibited by law.

Triad

a collaboration consisting of a sheriff, a police chief, and members of a senior citizen's group in the community who join together to reduce the victimization and better meet the needs of retired and other older persons. See *crime prevention.*

trial

court process in which allegations are examined and decided. Due to *plea bargaining,* most criminal cases do not go to trial. See *bench trial, jury trial.*

trial court

the lower court in which criminal cases are tried. Compare with *appeals court, supreme court.*

triangulation

a method of obtaining measurements at a crime scene. Triangulation, often recorded on graph paper, is accomplished by choosing two points other than the location of the object in question. Measurements from each point to the object are taken and transferred to the sketch. The location of the object is shown where the lines intersect on the sketch.

Trojanowicz, Robert

late professor of criminal justice at Michigan State University regarded by many as one of the founders, if not the father, of *community policing.* Trojanowicz, who developed a series of 12 principles for community policing, was also a fellow of the Kennedy School of Government at Harvard University.

truancy

the unexcused or unauthorized absence of a student from school. See *status offense, truancy.*

truant

a youth who is absent from school without authorization. See *status offender, truancy.*

true bill

the indictment document returned by a grand jury stating that a defendant has formally been accused of one or more offenses. See *bill of information, indictment.*

truncheon

a short, club-like weapon carried by a police officer. See *baton, billy club, nightstick, PR-24.*

trusty

a prison or jail inmate who is entrusted with duties or responsibilities requiring greater trust than that accorded other inmates. Typical trusty duties include chauffeuring automobiles for government officials and performing custodial duties.

trusty security

a level of security that permits greater freedom for the *trusty.* Compare with *close security, maximum security, minimum security.*

truth in sentencing

a phrase that conveys the intent to have convicted offenders serve the amount of jail or prison time to which they were sentenced. Truth in sentencing as a philosophy, expressed in numerous state laws enacted in the latter part of the 20th century, was a punitive response to the perception of victims and others that offenders were serving too little of the sentences given in court.

The Turner Diaries

a book written by William Pierce under the pen name of Andrew McDonald and published in 1980. *The Turner Diaries* is an example of modern neo-Nazi propaganda, espousing a revolution against the U.S. government, which has been protective and supportive of Jews and other targeted minorities.

twin studies

any criminological study employing identical twins to test for the inheritability of criminal tendencies. Twin studies enable researchers to examine the effects of the environment on twins reared separately.

Tyburn tree

an infamous gallows erected outside London in the 19th century. Originally designed for single hangings, the Tyburn tree was later modified to have three sides, each of which could accommodate several condemned offenders.

typology

a system of classifying things according to specified criteria. Typologies in criminology include those classifying offenders, crimes, victims, and theories of crime.

UCR

see *Uniform Crime Reports*.

Unabomber

the moniker given to *Theodore Kaczynski*.

unauthorized use of a motor vehicle

a criminal offense involving the taking of a motor vehicle, which is less serious than theft of a motor vehicle. Compare with *auto theft, carjacking, joyriding*.

underboss

in an organized crime family, the highest-ranking official just below the boss. Compare with *boss, consigliere*.

undercover

the state of infiltrating and investigating while concealing one's true identity. See *surveillance*.

undercover officer

a law enforcement officer who poses as a drug dealer or other criminal to investigate or infiltrate criminal enterprises.

undercover operation

a criminal investigation in which law enforcement officers pose as members of criminal organizations or otherwise disguise their true identities to gather detailed information about the organizations' inner workings. Undercover operations are common in narcotics investigations.

undercriminalization

the tendency of law to not treat seriously enough socially harmful behaviors.

Uniform Code of Military Justice

the body of criminal laws governing the behavior of armed services personnel in the United States.

Uniform Crime Reports (UCR)

a crime statistics program of the *Federal Bureau of Investigation (FBI)*. UCR has been the staple of national crime reporting since the 1930s. UCR collects summary-based information from law enforcement agencies on offenses reported and arrests made as well as more detailed data on homicides. The UCR is gradually being replace by the *National Incident Based Reporting System (NIBRS)*.

Unione Corse

a French organized criminal enterprise similar in nature to the *Mafia*.

United Bamboo

a Taiwanese crime organization.

unlawful assembly

the illegal gathering of persons in a public place.

unobtrusive measures

research techniques that do not involve contacting, interviewing, or otherwise inconveniencing research subjects. Unobtrusive measures include content analysis, the analysis of archival data, and certain observation techniques. The primary advantage of unobtrusive measures is that the researcher is less likely to bias or otherwise affect the data collection process.

unreported crime

crime not reported to law enforcement authorities. In addition to an amount of crime in general not reported, specific types of crime, such as sexual assault, tend to go unreported. See *dark figure of crime*.

uppers

Slang for a stimulant drug such as *amphetamine*. Compare with *downers*.

upperworld crime

criminal behavior by those in the upper socioeconomic strata of society. Examples of upperworld crime include *insider trading* and *embezzlement.*

Urban Institute

a private organization that conducts research and offers consulting services on a wide variety of criminal justice issues.

urinalysis

the chemical analysis of urine to check for the presence of drugs. When such a test yields a positive result, or what is known in criminal justice vernacular as *dirty urine,* it is indicative of drug use. Urinalysis is commonly used for substance abusers on *probation* or in treatment.

usury

lending money at an exorbitant interest rate. Although usury is generally associated with unlawful behavior, practices such as those employed by check-cashing businesses and similar loan establishments skirt the definition of usury, even though they have high lending rates and exploit low-socioeconomic patrons. Compare with *loan sharking.*

utilitarianism

the position in philosophy that happiness of the individual or of society in general is the desired end. See *classical school of criminology, Jeremy Bentham.*

vagrancy

not having a stable home or job. A charge for such a status. Many of those charged with vagrancy are suffering from mental illness. See *homeless persons, public order crime*.

vandalism

the malicious alteration or destruction of property, most often by juvenile offenders, often without obvious motive.

vehicular homicide

the unintentional killing of another through a motor vehicle accident.

vendetta

a series of revenge-motivated attacks. Vendettas are generally reprisals by organized criminals for a real or perceived wrong.

vengeance

the human desire to get even for a real or perceived injustice. Vengeance finds expression in the criminal law as *retribution*, one of several justifications for punishment. See *vigilantism*.

venue

the location where an offender is to be tried for a crime. See *change of venue*.

Vera Institute of Justice

a large, nonprofit organization in New York dedicated to improving the administration of justice through innovative research and practice.

vice squad
a police detective unit that specializes in investigating prostitution, gambling, drugs, and other so-called *vices*.

vices
moral violations, such as *prostitution*, gambling, or drugs. Vices are regarded by many as *victimless crimes*.

victim
an individual, business, or organization that suffers harm or loss as a result of a crime. See *secondary victim*.

victim advocate
a social worker who assists victims of crime to obtain services and navigate their way through the criminal justice process. Victim advocates, who may be former crime victims, refer victims to counseling, accompany them to court hearings, and sometimes participate in the preparation of victim impact statements. See *victim impact statement*.

victim assistance program
a program, often affiliated with a prosecutor's office, that refers crime victims to counseling and other needed services and helps them navigate their way through the criminal justice process. Victim assistance programs, which are staffed by victim advocates, can also be independent nonprofit agencies. See *victim advocate*.

victim compensation program
the formalized practice of paying a monetary sum from the government to crime victims, most often those who have suffered as a result of a violent offense, such as murder or physical or sexual assault. Victim compensation programs consider claims from victims who have filed an application, generally compensating victims of violent crimes. Federal funding for such programs comes from the *Office for Victims of Crime (OVC)*.

victim impact statement
a written or oral statement prepared and delivered prior to sentencing to inform a criminal court of the physical, emotional, financial, and other effects a crime has had on the victim, the family, friends, and the community. Victim impact statements are also prepared

for parole boards for the purpose of offering information about the offender's eligibility for parole.

victim notification

communication by state or local authorities that an offender is having a hearing or is about to be released. Victim notification is just one aspect of *victims' rights*.

victimization

the result of a crime in terms of human suffering or property loss.

victimization survey

a survey of citizens designed to determine the incidence and correlates of criminal *victimization*. Victimization surveys became popular with the recognition of the shortcomings of traditional crime statistics, such as the *Uniform Crime Reports (UCR)*. See *National Crime Victimization Survey (NCVS)*.

victimless crime

a type of crime in which there is no identifiable victim who has suffered harm or loss. *Prostitution* is often considered to be a victimless crime because both the prostitute and the *john*, or client, receive what they want: money, drugs, or some other gain to the prostitute and sex for the client. The victim of such a crime can be a third party, however, such as residents who live in a *red-light district* and experience a corresponding decrease in their property values or a less measurable sense of moral decay within the surrounding community. More serious crimes can sometimes accompany victimless crimes, such as armed robbery of johns by prostitutes or the assaulting of prostitutes by johns. See *overcriminalization*.

victim-offender mediation

efforts to bring together offender and victim to help each understand the perspectives and difficulties of the other. See *victim-offender reconciliation program (VORP)*.

victim-offender reconciliation program (VORP)

a program designed to bring victims and offenders together in an effort to gain mutual understanding.

victimology

the field of study that focuses on victims of crime and related issues.

victim-precipitated crime

a crime in which the victim plays a role in instigating the offender into a violent act. Critics of this concept maintain that it is the offender who engages in the wrongful behavior and therefore is responsible for his or her actions; research on violent crime such as homicide, however, suggests that victims do indeed sometimes engage in behavior that contributes to their own demise.

victims' bill of rights

a piece of legislation that ensures crime victims rights, such as the right to offer a *victim impact statement* in court and the right to notification of a convicted offender's release from prison.

victims' rights

a broad term that conveys concern for the ability of victims to more fully participate in the criminal justice system and to avoid further frustration and pain.

victim-witness program

a unit, often but not always housed within a prosecutor's office, whose staff serves the needs of crime victims by referring them to social services, providing emotional and other support during the court process, notifying victims of court hearings and offender release dates, and providing other services.

video piracy

the illegal reproduction, distribution, or sale of unauthorized copies of commercial videotapes.

vigilante

one who engages in *vigilantism*. Compare with *vengeance*.

vigilantism

the illegal practice by citizens of assuming the roles of police, prosecutor, judge, and sometimes even executioner in a misplaced attempt to seek justice.

violent crime

crimes against persons that involve bodily harm or the threat of bodily harm. Examples of violent crime are *murder, rape, robbery,* and *aggravated assault*. Compare with *property crime*.

Violent Criminal Apprehension Program (VICAP)
an initiative of the *Federal Bureau of Investigation (FBI)*. VICAP consists of a computerized database of detailed information on homicides, sexual assaults, and other violent crimes contributed by local law enforcement agencies. The VICAP computer is programmed to make daily searches for patterns that can identify possible crime series, enabling federal, state, and local investigators to avoid *linkage blindness*. The practical effectiveness of VICAP as a national investigatory tool for violent crime has been compromised by the unwillingness of local law enforcement agencies to regularly contribute data. See *National Center for the Analysis of Violent Crime (NCAVC)*.

voice stress analysis
an electronic means of analyzing voice patterns to detect deception.

voir dire
the process of selecting jurors prior to the commencement of a criminal trial. During voir dire, prospective jurors are questioned by both the *prosecutor* and the *defense attorney* to learn about their backgrounds and possible biases. See *jury selection*.

Volstead Act
federal legislation that prohibited the manufacture and sale of alcoholic beverages. The Volstead Act of 1919 prompted the beginning of *Prohibition*.

voluntary manslaughter
an intentional killing in circumstances that mitigate the seriousness of the crime to a lesser charge than murder. Such circumstances include provocations by the victim. Compare with *involuntary manslaughter*.

volunteers in probation (VIP)
a program in which lay volunteers assist probation officers in performing their duties and in meeting the needs of those on probation.

voyeurism
the practice of viewing people or materials in which they are depicted, often with little or no clothing, for the purpose of sexual gratification.

Walker spy ring

a spy ring that consisted of John Walker, his son, his brother, and his son's friend. The Walker spy ring passed information to the Soviets from the 1960s to the 1980s. It was determined that the information relayed by the Walker spy ring resulted in deaths of American operatives. See *espionage*.

Walnut Street Jail

a prison in Philadelphia in which prisoners spent their time alone in their cells. The Walnut Street Jail was illustrative of the *Pennsylvania system*.

war crimes

crimes committed during war that violate international law. Examples of war crimes are the rape of civilians, summary executions, and torture. War crimes are often tried in the International Court of Justice in The Hague, The Netherlands.

war on drugs

term used to describe the attempt by federal and state authorities to control the supply and distribution of illegal drugs. The war on drugs has drawn criticism for its emphasis on *supply reduction*, likened by many to *Prohibition*, which failed in its efforts to stem the flow of illegal alcohol. Compare with *demand reduction*. See *drug czar, Office of National Drug Control Policy (ONDCP)*.

warden

the administrator of a prison or other secure correctional facility.

warrant

a legal order issued by a judge that permits law enforcement authorities to arrest a suspect or search a specific location.

warrantless search

a search of a person, vehicle, or dwelling without a search warrant. Warrantless searches may be justified in certain circumstances.

Warren Commission

the federal commission convened by Earl Warren to investigate the assassination of President John F. Kennedy. The Warren Commission concluded that the assassination was the work of Lee Harvey Oswald operating alone. Many have since disputed this finding, with some proposing that Kennedy's assassination was the culmination of a vast conspiracy variously attributed to organized crime, Communists, or the *Central Intelligence Agency (CIA)*.

wartime trade violations

the prohibited sale or exchange of specific goods reserved for the war effort.

watch system

an early system of policing in which watchmen protected the walls and gates of the city and kept order among citizens. In contrast to modern police, those keeping peace under the watch system did not wear uniforms or carry firearms but instead rang bells at regular intervals. The increase in the consumption of alcohol, with its attendant violence, rendered the watch system obsolete.

Watergate

the 1972 break-in at the Democratic National Committee headquarters at the Watergate complex in Washington, D.C., and the ensuing attempts by the Richard M. Nixon administration to cover it up. The Watergate scandal led to the *indictment* and *conviction* of several conspirators and eventually to the resignation of President Nixon. See *political crime*.

weapons, carrying concealed

the illegal carrying of firearms, knives, or other dangerous ordnance prohibited by law. The carrying of concealed weapons is in fact legal in a number of jurisdictions. This has been a controversial issue between those who want to carry guns for self-protection and those who believe that such behavior will result in more violence and accidental shootings. See *affirmative defense*.

weapons, firing in public

the illegal discharge of firearms in a populated area where citizens would be at higher risk of being injured or killed.

weapons, selling

the illegal sale of weapons, often to ineligible buyers such as juveniles or convicted offenders.

Wedtech

a New York business and defense contractor that bilked the U.S. government out of millions of dollars from 1980-1985.

Wetterling Act

federal legislation established in 1994 that mandates that convicted sex offenders must register with local law enforcement agencies in the communities in which they live. Failure of sex offenders to comply with the provisions of the Wetterling Act can result in subsequent felony charges.

whistle-blower

an individual who reports wrongdoing in an organization of which he or she is personally knowledgeable. The discovery of *corporate crime* and other types of organizational deviance often depends on the willingness of someone to serve as a whistle-blower. Whistle-blowers face threats and other reprisals, despite federal and state laws intended to protect them.

white slavery

the illegal practice of selling or trading humans, often young females, generally for the purpose of forcing them to enter into *prostitution*. Also referred to as the *sex trade*.

white-collar crime

crimes committed by persons of high social status in the course of their occupations. The term was coined by American criminologist *Edwin H. Sutherland*. See *corporate crime, occupational crime, occupational deviance*.

Whitman, Charles

(1941-1966) a former U.S. Marine marksman who, in 1966, shot and killed 18 people and wounded 30 others on the campus of the University of Texas at Austin after killing his wife and mother.

Whitman, who was an engineering student at the university, was shot and killed by police when they stormed the library tower from which he wreaked this terror. See *mass murder*.

whodunit

a crime novel in which the murder suspect is not readily known. Whodunits often have several possible suspects. Generally, the identity of the perpetrator is not revealed until the end of the novel. See *mystery, suspense novel*.

whorl

a circle-like segment of a human *fingerprint*. See *arch, loop*.

Wickersham Commission

a federal commission convened in the 1930s to study problems in law enforcement and the administration of justice.

wife beating

the physical assault of a wife by her husband. See *domestic violence, husband beating, spousal abuse*.

wilderness camp

a camp in a remote wooded setting designed as a correctional alternative for delinquent youth.

Williams, Wayne

(1958–) a *serial killer* believed to be responsible for deaths of 28 young black men and boys in and around the city of Atlanta, Georgia, in the early 1980s. Williams, himself black, lured his victims with promises of fame. When the victims agreed to participate in homosexual acts, Williams killed them and dumped their bodies in various locations throughout Atlanta. He was sentenced to life imprisonment.

wiretapping

the practice of intercepting telephone transmissions for the purpose of gathering evidence for possible criminal prosecution. Wiretapping became an issue in the wake of the terrorist attacks on the World Trade Center and the Pentagon, when it was recognized that traditional wiretapping laws may no longer be sufficient to maintain national security against such threats.

withdrawal

the physical and psychological condition brought on by the discontinuation of drugs of *addiction*. In some cases, withdrawal can result in the death of the addicted patient. See *methadone maintenance*.

witness

a person who has personal, first-hand knowledge of a crime through the use of one or more of his or her senses, most often sight. Compare with *eyewitness*.

witness intimidation

the illegal act of threatening witnesses with harm. If successful, witness intimidation can result in dismissal of charges against the defendant. See *witness protection program*.

witness protection program

a federal program that permits witnesses testifying in dangerous cases to establish new identities in new locales to protect them and their families from possible reprisal. Those who take advantage of this program are typically witnesses who provide critical testimony in criminal cases against organized crime.

witness stand

a platform with a low wall behind it on which witnesses sit and testify in a courtroom. In Western courtrooms, the witness stand is located adjacent to the bench to the judge's left.

Wolfgang, Marvin E.

(1924-1998) American criminologist who spent most of his career at the University of Pennsylvania. Wolfgang is best known for pioneering studies of criminal homicide, the empirical assessment of crime seriousness, and measurement of delinquency over the life course. He was a student of criminologist *Thorsten Sellin*, with whom he worked for much of his career. Wolfgang served as president of the *American Society of Criminology (ASC)* and worked with prominent criminologists throughout the world, including Franco Ferracuti and Sir Leon Radzinowicz. See *Pennsylvania School of Criminology*.

Woolf Report

a report on access to justice in Britain prepared by Lord Woolf.

work release

a conditional release of an inmate from a correctional facility for the purpose of maintaining employment, obtaining training or education, or for other rehabilitative purposes. Those on work release must return to the facility at the end of the day.

workhouse

a local or regional correctional facility designed to house inmates for less than 1 year and in which inmates work as part of the correctional process. Compare with *jail*.

workplace violence

violent acts occurring at or near places of employment by employees or by their relatives and acquaintances. The problem of workplace violence, which gained national attention with shooting rampages by postal employees, has been the subject of specialized employee training.

Wournos, Aileen

a *serial killer* who murdered several men in Florida in the 1980s. Wuornos, a prostitute, shot and killed men who thought they were going to have sex with her. She claimed to suffer from a history of abuse.

wrongful conviction

the conviction in court of an accused person who, in fact, did not commit the alleged offense. A wrongful conviction can occur when a defendant, out of fear of a more severe *sentence*, pleads guilty to a crime of which he or she is not guilty. It can also occur as the result of erroneous *eyewitness identification*. The advent of *DNA testing* in criminal cases has led to the exoneration of a number of persons who were wrongfully convicted in the past. See *convicted innocent*.

wrongful execution

the execution of a person who did not commit the crime for which the penalty of death was eventually imposed. Despite the concern over this possibility, there is little evidence that many wrongful executions have occurred. See *convicted innocent, wrongful conviction*.

XYY chromosome

see *extra Y chromosome*.

Youngstown Gang

a notorious group of professional criminals who operated in Youngstown, Ohio, in the 1960s and 1970s. They committed a series of robberies, burglaries, and various other property crimes, all with the knowledge and support of not only organized crime in the areas but also local law enforcement.

Youth for Justice

an Ohio-based initiative to involve school-age youth in trying to reduce violence and other crime. Youth and their teachers discuss these issues in their schools and then share their solutions at a statewide conference in Columbus. See *law-related education*.

Youth Risk Behavior Survey (YRBS)

a biannual survey of high school youth about their dietary habits, substance abuse, fighting, and other behaviors that pose a risk to health. The YRBS is sponsored by the *Centers for Disease Control and Prevention (CDC)*.

Youth Service Bureau

an organization that offers prevention and intervention services for youth, especially those at risk of eventually entering the juvenile or criminal justice system.

Zebra killings

a series of homicides in California in the 1970s. Despite physical evidence left by the perpetrator and the testimony of a surviving victim, the killer was never caught. See *serial killer*.

zero tolerance

a policy in which law enforcement officials do not tolerate any disorder, especially public order crimes such as vagrancy, disorderly conduct, and soliciting for prostitution. Zero-tolerance policies were inspired by the *broken windows* position that minor incivilities can lead to increasing social disorder and crime if left unattended. Perhaps the best known example of zero tolerance is that of New York City, where Mayor Rudy Giuliani implemented a program wherein minor offenders were arrested and jailed, ultimately reducing the city's *crime rate*. See *arrest policies*.

Zionist Occupational Government (ZOG)

name given to the U.S. government by certain right-wing extremist groups. The name is intended to convey that the government is sympathetic toward Jews.

Zodiac killer

an unapprehended murderer responsible for a number of deaths in the San Francisco Bay area in the 1960s and 1970s.

zoophilia

sexual relations between a human and an animal. Also called *bestiality*.

GENERAL REFERENCES

American Psychiatric Association. (1994). *Diagnostic and statistical manual of mental disorders* (4th ed.). Washington, DC: Author.

Garland, D., & Sparks, R. (2000). *Criminology and social theory.* New York: Oxford University Press.

Geberth, V. J. (1996). *Practical homicide investigation: Tactics, procedures, and forensic techniques* (3rd ed.). New York: CRC Press.

Hagan, F. E. (1994). *Introduction to criminology* (3rd ed.). Chicago: Nelson-Hall.

Maxim, P. S., & Whitehead, P. C. (1998). *Explaining crime* (4th ed.). Boston: Butterworth-Heinemann.

McShane, M., & Williams, F. P., III. (1997). *Criminological theory.* New York: Garland.

Monk, R. C. (1996). *Taking sides: Clashing views on controversial issues in crime and criminology* (4th ed.). Guilford, CT: Dushkin.

Reid, S. T. (1997). *Crime and criminology* (8th ed.). Chicago: Brown & Benchmark.

Terrill, R. J. (1997). *World criminal justice systems: A survey* (3rd ed.). Cincinnati, OH: Anderson.

ABOUT THE AUTHOR

Mark S. Davis, PhD, is President of Justice Research & Advocacy, Incorporated, in Amherst, Ohio. He is former chief criminologist for the Ohio Office of Criminal Justice Services and the author of *Grantsmanship for Criminal Justice and Criminology* (2000). His research interests include violent crime, occupational deviance, and integrative theories of deviant behavior.